Lecture Notes in Earth Sciences

Lecture Notes in Earth Sciences

Edited by Somdev Bhattacharji, Gerald M. Friedman,
Horst J. Neugebauer and Adolf Seilacher

16

H. Wanner U. Siegenthaler (Eds.)

Long and Short Term Variability of Climate

Springer-Verlag
Berlin Heidelberg GmbH

Editors

PD Dr. Heinz Wanner
Universität Bern, Geographisches Institut
Hallerstrasse 12, CH-3012 Bern, Switzerland

PD Dr. Ulrich Siegenthaler
Universität Bern, Physikalisches Institut
Sidlerstrasse 5, CH-3012 Bern, Switzerland

ISBN 978-3-540-18843-8

Library of Congress Cataloging-in-Publication Data. Long and short term variability of climate /
H. Wanner, U. Siegenthaler, eds. p. cm.–(Lecture notes in earth sciences; 16) Papers presented at
a symposium held in Bern, Oct. 10–11, 1986, organized by the Swiss Commission for Climate and
Atmospheric Research. Includes index.
ISBN 978-3-540-18843-8   ISBN 978-3-540-38836-4 (eBook)
DOI 10.1007/978-3-540-38836-4

1. Climatic changes–
Congresses. I. Wanner, Heinz. II. Siegenthaler, U. (Ulrich), 1941-. III. Schweizerische Naturfor-
schende Gesellschaft. Schweizerische Kommission für Klima- und Atmosphärenforschung.
IV. Title: Variability of climate. V. Series.
QC981.8.C5L65 1988 551.6–dc 19 88-6542

© Springer-Verlag Berlin Heidelberg 1988

Originally published by Springer-Verlag Berlin Heidelberg New York in 1988

2132/3140-543210

Lake of Constance, winter 1830: The people of Rorschach enjoy walking on the frozen lake.

PREFACE

This volume includes papers presented at a symposium held at Bern on October 10 and 11, 1986, which was organized by the Swiss Commission for Climate and Atmospheric Research. It was the second symposium organized by this commission; the first one had been held also in Bern in 1983 (Das Klima, seine Veränderungen und Störungen, ed. C. Fröhlich; Birkhäuser Verlag, 1985). The symposium lectures met with a very good reception, and it appeared natural to publish them together in this Lecture Notes series. The authors were ready to submit their papers in camera-ready form, and we would like to thank them for their collaboration.

The symposium was planned mainly by H. Oeschger (then president of the Commission for Climate and Atmospheric Research), C. Fröhlich, G. Furrer and C. Pfister. R. Rickli, A.-C. Vogel-Clottu and E. Schüpbach were involved in the organization and administration of the conference. U. Neu assisted in the preparation of this book. We should like to express our thanks to all of them.

The Swiss Academy of Sciences made the symposium possible by providing the necessary financial support.

Bern, December 1987   H. Wanner
                                       U. Siegenthaler

CONTENTS

INTRODUCTION

The awareness that mankind is able to influence and modify not only the local but also the global climate has led to a strongly growing interest in climate research. Strengthened research activities, which also made use of improved and novel experimental techniques, have yielded a wealth of information on climatic patterns in the past. At the same time, climate modelling has made much progress. While some questions have been answered, new problems have been recognized. One question related to anthropogenic climatic change is about the nature and causes of natural variations, against the background of which man-made changes must be viewed.

The contributions to this volume all deal with the variability of climate. Some papers are reviews of the knowledge to a current topic, others have more the character of an original contribution. The observational studies cover the range from year-to-year variations up to glacial-interglacial contrast, thereby going from instrumental data to results from proxy records.

The question whether the sun's energy output varies with time has long been considered, but reliable data only became available when measurements could be undertaken outside of the atmosphere. The paper by C. Fröhlich deals with the short and longer-term variability of the solar "constant". The measurements are carried out with absolute radiometers operating on board of balloons, rockets or spacecraft. By using satellite data obtained with high-precision radiometers as well as spot measurements it has been possible to detect the first unequivocal evidence of a long-term trend of the solar constant. For the period since 1980 it shows a decrease of -0.019 % per year.

P.D. Jones and P.M. Kelly calculated mean annual hemispheric temperatures since 1861 and discuss possible causes for the temperature variations. They find that variations in solar activity, increasing carbon dioxide, volcanic activity and the El Niño / Southern Oscillation (ENSO) phenomenon are the most probable causes of variations in global mean temperature on the 1 to 100 year time scale. Large explosive volcanic eruptions as well as the ENSO phenomenon both have similar effects, of the order of 0.1 to 0.2 °C, on hemispheric temperature. The duration of the maximum effect is of the order of 6 months and it occurs at some time within the two years immediately after the volcanic eruption or warm or cold event. The two factors are responsible for between 30 and 50 % of the interannual variability in the hemispheric temperature records.

A critical insight into the possibilities and problems of dendroclimatology is given by F. Schweingruber. He shows that site factors exert a stronger influence on growth ring formation than regional conditions. For that reason, great attention is being paid to the selection of sites and individual trees. The author emphasizes that abrupt growth changes persisting for more than three years incorporate climatic signals like temperature and precipitation anomalies. By using a set of coloured maps he compares the spatial pattern of the anomalies of temperature (July - September), maximum density and tree-ring width for Europe. The coincidence between temperature and maximum density anomalies is obvious, but ring width anomaly patterns cannot be explained by the influence of one climatic factor alone. Schweingruber demonstrates that in Switzerland a phase of clear growth reduction can be observed between 1945 and 1954. He also indicates that the growth patterns are strongly governed by deficits in summer precipitation. The strong irregularities of the last years may well open a large field for dendroclimatological research within the next years.

C. Pfister investigates the weather patterns of the vegetation period between 1270 and 1425 and compares tree-ring and grape harvest records of this warm period with corresponding data until the end of the "Little Ice Age". Although continuous proxy data of the warm period of the High Middle Ages are not yet available, he can show that positive anomalies occurred more than once every decade between 1269 and 1339. After 1400 they became very rare, and not a single occurrence is measured for the seventeenth century. A "climatic watershed" is marked by the early fourteenth century. The shift from the warm climate of the High Middle Age to the "Little Ice Age" took only about two decades and was characterized by an enormous variability. One of the warmest years within the last millenium was certainly 1420 when the wine harvest in Western and Central Europe began at the end of August and was advanced by a month compared with the long-term average for Western Europe! In view of the strong anomalies reported here for earlier centuries, the question arises whether the weather excursions Central Europe has witnessed in the past few years were induced by human impact or just represent natural fluctuations.

The subject of the paper by J.C. Duplessy, L. Labeyrie and P.L. Blanc is the history of the glacial-interglacial cycles, about which much has been learned form ocean sediment studies. The oxygen isotope ratio $^{18}O/^{16}O$ in ocean water was lower during the ice age than at present, because large amounts of water with low $^{18}O/^{16}O$ ratio had been transferred from the ocean to the continental ice. This is recorded in the oxygen isotope ratios of deep-sea sediments, which, however, were modified by temperature changes affecting the isotope fractionation between sea water and carbonate. The French authors now have been able to disentangle the effects of ice volume and temperature by careful analysis of sediment cores from the Norwegian Sea, using the fact that there, the deep-water temperature is near the freezing point today and can therefore not have been significantly lower during glacial time. They present a table with the oxygen isotope ratio of mean ocean water as a function of time for the last 135,000 years that will be useful for isotope stratigraphic studies.

The part of the book dealing with climate modelling includes three papers. H. Grassl discusses basic applications, the main problems and some important results of numerical models of climate. He distincts meteorological and climate applications and points to the three principal sources of errors, which are numerical and parameterization errors as well as errors caused by incomplete equations. Although modellers are well aware that a climate model should include all relevant compartments of the climate system, a tested coupled ocean-atmosphere model is still not available, because the complexity of such a model still exceeds the available computing resources. Strong efforts have been made to answer the question how the climate system will react to the changing composition of the atmosphere. While modellers agree that a doubling of the CO_2 content will lead to a mean global temperature increase between 1.5 and 4.5 K, it is much more difficult to estimate how chemical processes (e.g. photochemical reactions) and aerosol particles modify the climate and its components.

The astronomical, or Milankovitch, theory of the ice ages, according to which the variations in the geometry of the earth's orbit are the fundamental cause of the glacial-interglacial cycles, is now widely accepted. Generally, the variation of the insolation at the top of the atmosphere has been considered so far. C. Tricot and A. Berger consider instead the radiation absorbed at the earth's surface, which is more relevant for climate. Tricot and Berger have computed the absorbed radiation for different latitudes for the past 200,000 years and discuss the results in their contribution.

Finally, U. Siegenthaler discusses the implication of measurements made on air occluded in polar ice cores. They have shown that the atmospheric concentration of CO_2 was about 30 percent lower during the ice age than during the Holocene. Climate model studies indicate that the lower CO_2 level significantly contributed to the cold ice age temperatures. The model result that cooling in the Southern Hemisphere may have been largely a result of the CO_2 variations is of special importance; this would explain why glaciations were synchronous in both hemispheres, a fact that is hard to understand from the Milankovitch theory of climate alone. The cause of the CO_2 variations must have been changes in the ocean which, however, may have been initiated by climatic events. Thus, it appears necessary for understanding the glacial cycles to consider climate and the carbon cycle simultaneously as one interactive complex climate-CO_2 system.

Obviously, the papers in this book do not represent a complete survey. We think, however, that they constitute a collection of articles dealing with a number of topical issues and thus illustrate very well the problems climate research is presently faced with. We hope that this volume may provide an insight into some important contributions of current research in Europe on long and short term variability of climate.

Bern, December 1987

U. Siegenthaler   H. Wanner

VARIABILITY OF THE SOLAR "CONSTANT"

C.Fröhlich
Physikalisch-Meteorologisches Observatorium
World Radiation Center
CH-7260 Davos Dorf, Switzerland

1. Introduction

Since the first clear evidence of changes in the solar "constant" S_0 from the records of the Active Cavity Radiometer for Irradiance Monitoring (ACRIM, Willson, 1979) on the Solar Maximum Mission (SMM) and of the Hickey-Frieden radiometer (Hickey et al, 1980) on NIMBUS 7 proving that the sun is indeed a "variable" star, the interest on solar irradiance variability on all time scales has very much increased (Willson, 1984; Fröhlich, 1987). Atmospheric physicists and climatologists are concerned, because of possible effects on the earth's energy balance. Solar physicists, on the other hand, became interested, because global changes of the solar output have been doubted for a long time and their reality obviously leads to some revision of the understanding of the behaviour of the sun.

2. Solar Irradiance Measurements

The solar "constant" is the solar irradiance at 1 astronomical unit (1 A.U.= mean sun-earth distance) integrated over the whole spectrum. Instruments for the accurate measurement of this quantity are the so-called absolute radiometers (e.g. Kendall et al., 1970; Geist, 1972; Willson, 1979; Brusa et al., 1986) which are also used as reference instruments for the calibration of operational radiometers in meteorological networks. They are all based on the measurement of a heat flux through an electrically calibrated heat flux transducer. The radiation is absorbed in a cavity which ensures a high absorptivity (typically >99.95%) over the spectral range of interest for solar radiometry (200 nm - 10 μm). The heat flux transducer consists of a thermal impedance and of thermometers (e.g. thermopile, resistors) to sense the temperature difference across it. Heat developed in the cavity is conducted to the heat sink of the instrument and the resulting temperature difference across the thermal impedance is sensed. The sensitivity of the heat flux transducer is calibrated by shading the cavity and measuring the temperature difference while dissipating a known amount of electrical power in a heater element which is mounted inside the cavity. In the so-called active mode of operation an electronic circuit maintains the temperature signal constant by accordingly controlling the power fed to the cavity heater - independent of the mode, that is whether the cavity is shaded or irradiated. The substituted radiative power is then equal to the difference in electrical power as measured during the shaded and irradiated periods respectively. In the ideal case of a perfect substitution of radiative by electrical power, the irradiance S would simply be:

$$S = (P_s - P_i)/A$$

where P_s and P_i is the electrical power dissipated with the cavity shaded and irradiated respectively, and A is the area of the detector. However, there are many deviations from this ideal behaviour and the $1/A$ term will have to be replaced by a more elaborate expression accounting for these effects. The process of experimentally determining the size of these effects is called experimental characterization (Brusa et al., 1986). The uncertainty of the characterization determines the absolute accuracy of the radiometer which is of the order of ±0.2% for present state-of-the-art solar radiometry.

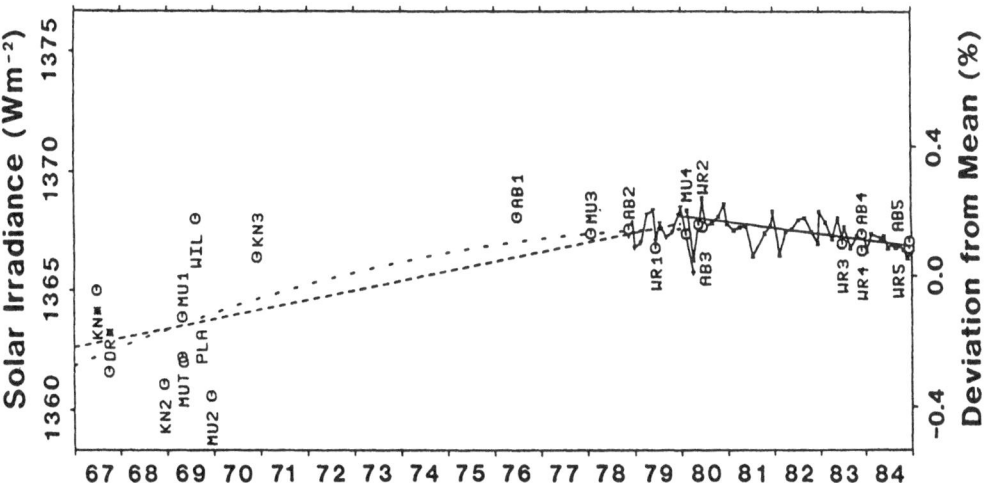

Figure 1: Measured values of total solar irradiance 1967 to 1983 (for the labels see text). The full curve labeled with crosses represent the result of the satellite measurements (one data point every month) for 1978 to 1980 from NIMBUS-7 (Hickey et al, 1982) and for 1980 to 1985 from SMM/ACRIM (Willson et al, 1986). For the discussion of the trends see section 4.

The solar radiation is depleted in the earth's atmosphere by absorption and scattering, which depends strongly on the wavelength. Thus accurate determinations of S_0 can only be made from high altitude balloons (above 35 km), rockets or spacecrafts. Determinations of S_0 from mountain tops were performed by the Smithsonian Institution under the leadership of Abbot (e.g. Abbot, 1942) continuing the pioneering work of Langley. Although sophisticated methods were applied to correct for the atmospheric extinction the results only marginally revealed the small solar "constant" variations (e.g. Foukal et al., 1977, Hoyt, 1979). Direct measurements from balloons, rockets and spacecrafts started in the late sixties and have been continued to present with a gap between 1971 and 76. The results are shown in Fig.1: the Soviet balloon flights KN*, KN2 and KN3 (Kondratyev & Nikolsky, 1970, 1979), the X-15 rocket airplane flight DR* (Drummond et al, 1968), radiometry on the Mariner VI and VII spacecraft PLA (Plamondon, 1969), the balloon flights of the Denver University group MU1 to MU4 (Murcray et al, 1969; Kosters & Murcray, 1981), the balloon flight WIL of Willson (1973), the NASA calibration rocket flights AB1, AB2, and AB3 (e.g. Willson, 1981), the PMOD/WRC balloon flights WR1, WR2, and WR3 (Brusa, 1983) and the spacecraft measurements on NIMBUS 7 (Hickey et al, 1982) and on SMM (Willson, 1984). These data are supplemented by the results from two rocket flights with PMO and ACR instruments: WR4 & 5 and AB4 & 5, and updated data from SMM (Willson et al, 1986). This summary demonstrates the improvements achieved in absolute radiometry especially since 1980. The scatter between the individual results in the late sixties is mostly instrumental, whereas the variability after 1980 is mostly of solar origin.

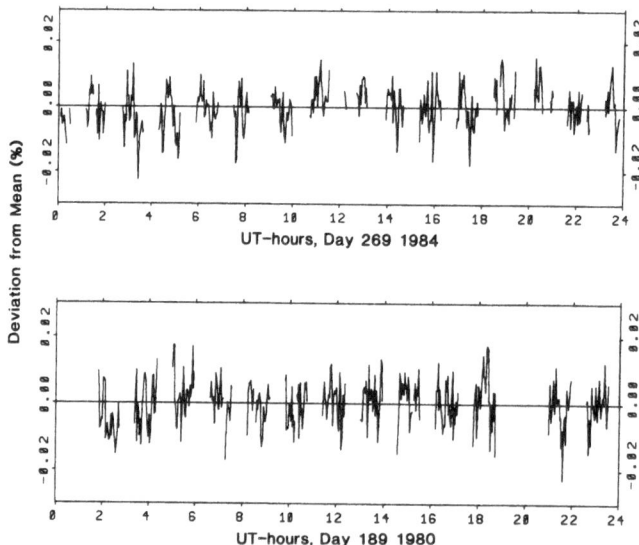

Figure 2: SMM/ACRIM individual solar irradiance measurements (every 131 s) during one day in 1980 (lower panel) and 1984 (upper panel). The periodically missing data are due to the modulation by the spacecraft orbit around the earth with a period of 94 to 96 minutes.

3. Variability of the Solar "Constant"

The solar irradiance variability on time scales from a few minutes to several months is illustrated by the time series shown in Fig.2 to 4 for two different periods during the solar activity cycle: 1980 (lower panels) around the maximum and 1984 (upper panels) close to the minimum of the solar cycle 21. Fig.2 shows the variability during one

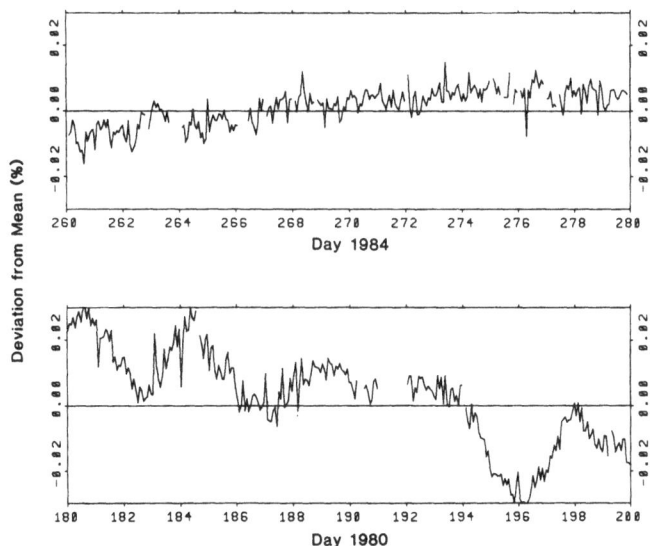

Figure 3: SMM/ACRIM solar irradiance data (orbital means) during 20 days in 1980 (lower panel) and 1984 (upper panel).

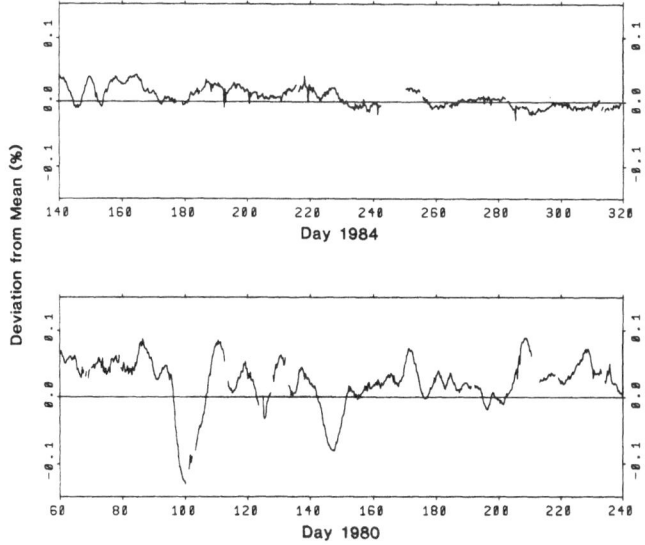

Figure 4: SMM/ACRIM solar irradiance data (orbital means) during 180 days in 1980 (lower panel) and 1984 (upper panel).

day. These short-term variances have a mean peak-to-peak (p-p) amplitude of about 200 ppm (parts per million) and are very similar during
both periods. For the 20 days period shown in Fig.3, however, the
behaviour in 1980 is quite different from the one in 1984 with p-p
variations of 0.06% and 0.03% respectively. Also the main periods of
the variability are quite different. This is even more pronounced for
the period of 180 days shown in Fig.4: the 1984 variance remains at the
same level whereas the 1980 variance reaches several tenths of a percent. The short-term variations of Fig.2 are mainly due to solar pressure oscillations (e.g. Woodard, 1984, Fröhlich et al., 1984) and
partly due to granulation. Some of the variability of Fig.3 may be
caused by internal gravity oscillations with periods from several hours
to days (e.g. Fröhlich et al., 1984; Fröhlich, 1986). The following
discussion will mainly concentrate on the variations shown as time
series in Fig.4 and on the trends indicated in Fig.1.

The variability of the solar irradiance on time scales of days has
been discussed by several authors (e.g. Willson et al, 1981; Hickey et
al, 1982) and several models have been established for the explanation
of the variance mainly by sunspot blocking and facular enhancement
(e.g. Hudson et al, 1982; Schatten et al, 1982; Hoyt & Eddy, 1983;
Foukal & Lean, 1986; Pap, 1986). Most of these models are tested
against the records of ACRIM/SMM and of H-F/NIMBUS-7. One issue in this
context is the question whether the energy blocked by the sunspots is
immediately balanced by the emissions in faculae (e.g. Chapman, 1984)
or whether the blocked energy has to be stored below the active regions
and emerges only slowly over periods of months or years (e.g. Foukal et
al., 1983). Even if the energy were exactly balanced, the irradiance at
1 A.U. would still vary because of the different spatial distribution
on the solar surface and the different angular emission pattern of the
two features. Recent results indicate that the facular contribution to
S_0 is at least comparable to that of spots, when integrated over months
(Foukal & Lean, 1986). This issue is very important for our understanding of the behaviour of active regions and for adequately modelling the
solar irradiance modulation which in turn is needed to understand
climate changes forced by solar variability.

Fig. 5 shows the power spectra of ACRIM/SMM data in the frequency
range up to 10 μHz (11.6 μHz corresponds to a period of 1 day) for 1980
and 1984. These measurement periods are before failure and after repair
of the accurate pointing system of the SMM spacecraft and cover 9 and 8
months respectively. The difference in the spectra is mainly due to the
difference of the activity level of the sun during these periods. The
two major peaks at low frequencies with periods of 51.4 and 23.5 days
are reduced by more than a factor of ten to a broad peak centered
around a 17-days period in 1984 (half power points at periods of 46.3
and 10.6 days respectively). The period of 51.4 days is also found in
the occurrence of high energy flares (Rieger et al, 1984), in the
Zürich sunspot number and in solar diameter data (Delache et al, 1985).
Although the power spectrum of the projected sunspot area in 1980 shows
a significant peak at 27 days, the peak in irradiance is shifted to

23.5 days. Cross-spectral analysis of the two spectra also reveals a very weak coherence between irradiance and sunspot area at 27 days (Fröhlich, 1984; Foukal & Lean, 1986). Furthermore, the phase between the signals from sunspots and irradiance at 27 days indicates that it is more likely an enhancement which could be due to faculae than a depletion by spots. Obviously, the differences in spatial distribution of spot and faculae on the solar surface and their different evolution in time make that the individual contributions to the total irradiance signal can no longer be distinguished. The depletion of the irradiance due to sunspot blocking seems also to depend on the age of the spot and not only on its projected area; young and active spots have a stronger influence than old and passive spots and indeed a frequency analysis of the evolution of young and active spots in 1980 shows the same period of 23.5 days as the ACRIM/SMM irradiance (Pap, 1986). Other significant peaks in the spectra are found at 7.0, 4.8, 3.4 and 1.3 days. The 4.8 and 1.3 days periods are found in the spectra of both years. In the 1984 spectrum also many significant peaks between 5 and 9 μHz similar to the 1.3 days peak are found, the origin of which is still unknown.

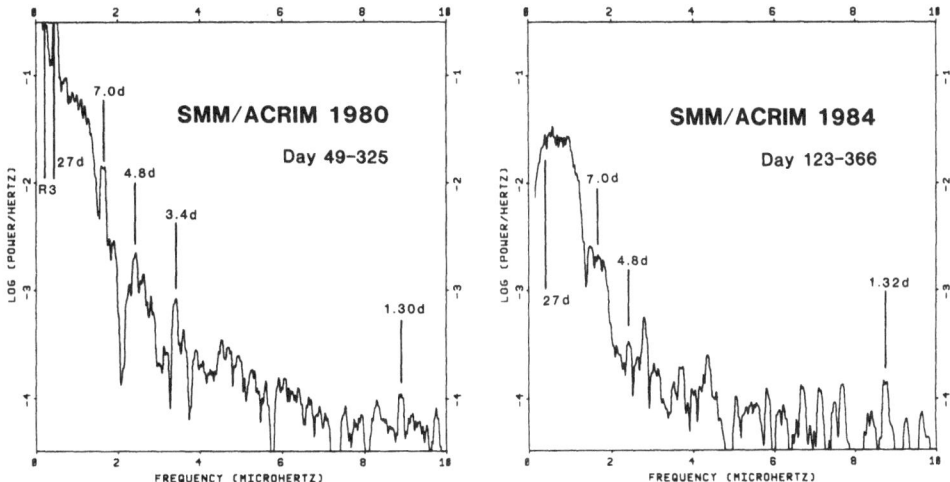

Figure 5: Comparison of power spectra of ACRIM irradiance data during 1980 (277 days, left panel) and 1984 (244 days, right panel). The label R3 refers to Delache et al., 1985, and is a period found in the occurence of flares (Rieger et al., 1984). Note the lack of a 27-days peak present in power spectrum of the 1980 sunspot data.

Table 1 summarizes the distribution of the variance in the power spectra of 1980 and 1984. Most of the variance is concentrated in the range below about 2 μHz (more than 97% in 1980 and 92% in 1984) and it is also here where the biggest change in variance by nearly a factor of 7 (2.6 in amplitude) from 1980 to 1984 occurs. In the range from 2 to 5.8 μHz the amount is less than 1% of the total variance and also the change is much smaller (factor of 2.8 in variance and 1.7 in ampli-tude). Above 5.8 μHz the variance is for both years very small relative to the total and of the same magnitude for both years. In summary the solar activity influences the variance of the total irradiance significantly, especially at low frequencies.

Table 1: Variance of Solar Irradiance for the frequency range from 0.1 to 80 μHz in 1980 and 1984.

Range Frequency μHz	Period days	Variance ppm² 1980	1984	Standard Deviation ppm 1980	1984
0.1 - 80	0.14 - 110	177000	27200	421	164
0.1 - 2.1	5.6 - 110	172000	24900	416	158
2.1 - 5.8	2.0 - 5.6	1480	518	38.5	22.8
5.8 - 10	1.2 - 2.0	276	236	16.6	15.4
10 - 40	0.29 - 1.2	791	770	28.1	27.7
40 - 80	0.14 - 0.3	535	635	23.1	25.2

4. Long-term Trends

Trends purportedly found in the early measurements of S_0 by the Smithsonian Institution were generally doubted on the basis of the large atmospheric corrections involved. Determinations, made occasionally from aircraft, balloons, X-15 rocket aircraft, and mariner satellites in the late 1960's seemed also too uncertain in both calibration and intrinsic error to allow comment on real variations in S_0 during that period. The modern satellite data together with spot measurements from sounding rockets and balloons, however, allow for the first time to assess confidently possible trends in S_0.

Critical reviews of measurements of S_0 made after 1967 have been given elsewhere (Fröhlich, 1977; Fröhlich & Eddy, 1984; Fröhlich, 1987). Most of the earlier values have been adjusted from original published values to conform to a common standard, the World Radiometric Reference (WRR). This was done in the manner described earlier by Fröhlich (1977). In addition, to insure uniformity the atmospheric correction for all balloon measurements was recomputed using the scheme adopted in the reduction of the PMOD/WRC results (Brusa, 1983). The results of the 1980 experiment of the University of Denver (MU4) can be directly compared with the results of the rocket experiments AB2 and AB3 and the balloon fights WR1 and WR2 using the NIMBUS 7 record for interpolation. Thus an absolute value can be attributed to MU4 independent of atmospheric transmission correction. As MU1 in 1969 and MU4 in 1980 were carried out at the same altitude and with the same instrument, the calibration for MU4 can be transferred to MU1 making use of the difference of 0.38 per cent between the two determinations reported

by Kosters and Murcray (1981). The result is labeled MUT in Fig.1. The
close agreement between MU1, MU2 and MUT demonstrates the stability of
the Denver instrumentation and supports the upward trend.

   The linear regression analysis to the spot measurements before
1981 shown in Fig. 1 suggests an increase of the solar constant until
1980 at a rate of 0.029 per cent per year. This trend is significantly
different from zero at the 99.9 per cent confidence level. It is of the
same sign as the change of 0.38 per cent between 1969 and 1980 noted by
Kosters and Murcray, 1981, although the slope is only about three-
quarters as great. Higher-order analysis gives an improved fit to the
composite data, shown as the curved line in Fig. 1 and indicating a
maximum around 1979. One must bear in mind, however, that most of the
data taken in the early part of the set were the results of inherently
less-reliable balloon measurements which could be influenced by a
common systematic overestimation of the stratospheric transmittance. In
this case, one would have to assume either an anomalous (high) concentration of stratospheric ozone - about 1.5 times the climatological
value - or an increased opacity due to an enhanced abundance of high-
altitude aerosol. The latter might ensue from a major volcanic eruption, although there was none reported in this period. All this seems

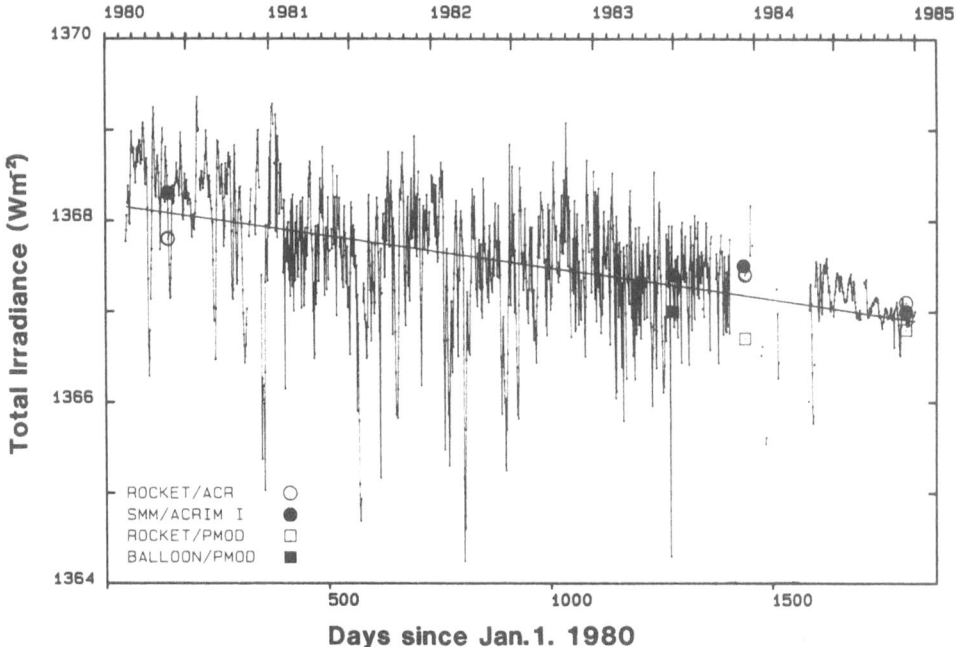

Figure 6: Time series of SMM/ACRIM daily mean results for the period
    from 1980 to 1985. The linear least square fit shown has a slope
    of -0.019% per year. Independent total irradiance observations by
    sounding rocket and balloon experiments show good agreement with
    ACRIM results (from Willson et al., 1986).

unlikely (see also Kosters & Murcray, 1981). Furthermore, the results from Mariner and the X-15 should be exempt from atmospheric effects and they support the lower values of the early balloon measurements. Thus it is concluded that the low values of S_0 from the late 1960's are most probably real.

For the period since 1980 the ACRIM data have been used to determine the trend as shown in Fig.6. A full discussion of this result is given by Willson et al., 1986. The linear fit for this period is calculated from the daily means of the ACRIM data and yields a trend of 0.019 per cent per year. This trend is confirmed by the NIMBUS 7 data and the spot measurements during this period and is the first clear evidence of a long-term trend of the solar constant.

The extant measurements of S_0 from 1967 to 1985 suggest a slow oscillation in absolute value which could be part of a 22-year modulation with a peak-to-peak amplitude of about 0.4 per cent coincident with the magnetic cycle of the sun. Due to the missing data between 1971 and 1976, however, it is not clear whether the trend between 1969 and 1980 was continuous or had a dip during the minimum. Even if the latter would be the case the lower data in 1969 could still be explained by the fact that the activity maximum in 1969 was only about two thirds of the strength of the one in 1980.

5. Conclusions

The power of the irradiance variability spectrum from about 100 nHz (110 days) to 80 μHz (3.5 hours) can be divided into three major domains with the following characteristics:

- From 100 nHz to 2 μHz (5.8 to 110 days) the spectrum is dominated by solar activity the power of which changes during the course of the solar cycle by up to one order of magnitude. Moreover, the spectrum is characterized by prominent peaks at periods of 51.4, 23.5, 7.0 and 4.8 days. The variance in this range amounts to 172000 and 24400 ppm^2 for 1980 and 1984 respectively.

- From 2 to 15 μHz (18.5 hours to 5.8 days) the spectrum follows a $1/\nu^2$ law, which may be partly due to internal gravity modes. The variance in this range is 1915 and 908 ppm^2 for 1980 and 1984 respectively.

- From 15 μHz to 80 μHz (18.5 to 3.5 hours) the spectrum follows a $1/\nu$ law, which may be partly due to instrumental noise. The variance in this range is 1200 ppm^2 for both years.

As to the long-term changes, trends of the order of 0.02 per cent per year do exist. The question whether the up and down trends with a peak around 1980 belong to an oscillatory modulation of the solar output with a period of 11 or 22 years can only be answered in the future. The decrease of S_0 from high to low activity could be due to the same mechanism in the sun which produced the "little ice age" in the 17th century in Europe, when the solar activity was very low over a period of many solar cycles (e.g. Eddy, 1977). The question of the 11 and/or 22 years modulation of the solar output is also very important in the context of the results of analysis of ancient varves (e.g. Williams & Sonett, 1985; Sonett & Trebisky, 1986) which indicate also a modulation of the climate with a major period of 11 years and a minor one of 22 years. Moreover the ratio of the 11/22 years modulation amplitude seems to decrease with time. As the answer is not only important for the interpretation of changes of the earth climate but also for the understanding of the sun itself, monitoring of the solar "constant" has to be continued.

Acknowledgements. I thank R.C.Willson, Jet Propulsion Laboratory, Pasadena, U.S.A., for providing unpublished ACRIM data and for many helpful discussions. Acknowledgements are extended to the Swiss National Science Foundation for their continuous support of this work at PMOD/WRC.

REFERENCES

Abbot, C. 1942: Revised Results of Solar Constant Observing 1923 to 1939, *Ann. Smithson.Astrophys.Observ.*, **6**, 83.

Brusa, R.W. 1983: Solar Radiometry, *Dissertation ETH No. 7181*, Zurich.

Brusa, R.W. & Fröhlich, C. 1986: Absolute Radiometers (PM06) and their Experimental Characterization, *Appl. Opt.*, **25**, 4173.

Chapman, G.A. 1984: On the Energy Balance of Solar Active Regions, *Nature*, **308**, 252.

Delache P., Laclare, F. & Sadsaoud, H. 1985: Long Period Oscillations in Solar Diameter Measurements, *Nature*, **317**, 416.

Drummond, A.J., Hickey, J.R., Scholes, W.J. & Laue, E.G. 1968: New Value of the Solar Constant of Radiation, *Nature*, **218**, 259.

Eddy, J.A. 1977: Climate and the Changing Sun, *Clim. Change*, **1**, 173.

Foukal, P., Mack, P.E. & Vernazza, J.E. 1977: Effect of Sunspots and Faculae on the Solar Constant, *Astroph.J.*, **215**, 952.

Foukal, P., Fowler, L.A. & Livshits, M. 1983: A Thermal Model of Sunspot Influence on Solar Luminosity, *Astroph.J.*, **267**, 863.

Foukal, P. & Lean, J. 1986: The Influence of Faculae on Total Solar Irradiance and Luminosity, *Astroph.J.*, **302**, 826.

Fröhlich, C. 1977: Contemporary Measurements of the Solar Constant, in *"The Solar Output and Its Variation,"* ed. O.R.White, Colorado Associated University Press, Boulder, p.93.

Fröhlich, C. 1984: Solar Variability for Periods of Days to Months, *Adv. Space Res.*, **4**, No.8, 117.

Fröhlich, C. & Delache, P. 1984: Solar Gravity Modes from ACRIM/SMM Irradiance Data, in *"Solar Seismology from Space"*, ed. R.K.Ulrich,

JPL Publ.84-84, Pasadena, CA., 173.

Fröhlich, C. & Eddy, J.A. 1984: Observed Relation between Solar Luminosity and Radius, *Adv.Space Res.*, **4**, No.8, 121.

Fröhlich, C. 1986: Solar Gravity Modes from ACRIM/SMM Irradiance Data, *Advances in Helio and Astroseismology, IAU Symposium 123*, Aarhus.

Fröhlich, C. 1987: Variability of the Solar "Constant" on Time Scales of Minutes to Years, *J.Geophys.Res*, **92**, **D1**, 796.

Geist, J. 1972: Fundamental Principles of Absolute Radiometry and the Philosophy of this NBS Program (1968-1971), *Natl.Bur.Stand.U.S. Tech.Note*, **5941**.

Hickey, J.R., Pellegrino, P., Mashhoff, R.H., House, F. & Vonder Haar, T.H. 1980: Initial Solar Irradiance Determination from NIMBUS 7 Cavity Radiometer Measurements, *Science*, **208**, 281.

Hickey, J.R., Alton, B.M., Griffin, F.J., Jacobowitz, B., Pellegrino, P. & Smith, E.A. 1982: Observations of the Solar Constant and its Variations Emphasis on NIMBUS 7 Results, in *"Proc. IAMAP Rad. Comm. 3rd Scientific Assembly, Hamburg 1981"*, NCAR, Boulder.

Hoyt, D.V. & Eddy, J.A. 1983: Solar Irradiance Modulation by Active Regions from 1969 through 1981, *Geoph.Res.Letters*, **10**, 509.

Hoyt, D.V. 1979: The Smithonian Astrophysical Observatory Solar Constant Program, *Rev.Geophys.Space Phys.*, **17**, 427.

Hudson, H.S., Silva, S., Woodard, M. & Willson, R.C. 1982: The Effect of Sunspots on Solar Irradiance, *Solar Phys.*, **76**, 211.

Kendall, J.M. & Berdahl, C.M. 1970: Two Blackbody Radiometers of High Accuracy, *Appl.Opt.*, **9**, 1082.

Kondratyev, K.Y. & Nikolsky, G.A. 1970: Solar Radiation and Solar Activity, *Quart.J.Roy.Meteor.Soc.*, **96**, 509.

Kondratyev, K.Y. & Nikolsky, G.A. 1979: The Stratospheric Mechanism of Solar and Anthropogenic Influences on Climate, in *"Solar Terrestrial Influences on Weather and Climate"*, ed. B.M.McCormac & T.A.Seliga, Reidel, Dordrecht, Holland, p.317.

Kosters, J.J. & Murcray, D.G. 1981: Change in the Solar Constant between 1968 and 1978, in *"Variations of the Solar Constant"*, ed. S.Sofia, NASA Report CP-2191.

Murcray, D.G., Kyle, T.G., Kosters, J.J. & Gast, P.R. 1969: The Measurements of the Solar Constant from High Altitude Balloons, *Tellus*, **XXI**, 620.

Pap, J. 1986: Variation of the Solar Constant during the Solar Cycle, *Astrophys.Space Sci.*, **127**, 55.

Plamondon, J.A. 1969: The Mariner Mars 1969 Temperature Control Flux Monitor, *JPL Space Science Program Summary* 3, 162.

Rieger, E., Share, G.H., Forrest, D.J., Kanbach, G., Reppin, C. & Chupp, E.L. 1984: A 154-day Periodicity in the Occurrence of Hard Solar Flares?, *Nature*, **312**, 623.

Schatten, K.H., Miller, N., Sofia, S. & Oster, L. 1982: Solar Irradiance Modulation by Active Regions from 1969 through 1980, *Geoph. Res. Letters*, **9**, 49.

Sonett, C.P. & Trebisky, T.J. 1986: Secular Change in Solar Activity derived from Ancient Varves and the Sunspot Index, *Nature*, **322**, 615.

Williams, G.E. & Sonett, C.P. 1985: Solar Signature in Sedimentary Cycles from the late Precambrian Elatina Formation, Australia, *Nature*, **318**, 523.

Willson, R.C. 1973: New Radiometric Techniques and Solar Constant Measurements, *Solar Energy*, **14**, 203.

Willson, R.C. 1979: Active Cavity Radiometer Type IV, *Appl.Opt.*, **18**, 179.

Willson, R.C. 1981: Solar Total Irradiance Observations by Active Cavity Radiometer, *Solar Physics*, **74**, 217.

Willson, R.C. 1984: Measurements of Solar Total Irradiance and its Variability, *Space Science Rev.*, **38**, 203.

Willson, R.C., Gulkis, S., Janssen, M., Hudson, H.S. & Chapman, G.A. 1981: Observations of Solar Irradiance Variability, *Science*, **211**, 700.

Willson, R.C., Hudson, H.S., Fröhlich, C. & Brusa, R.W. 1986: Observation of a Long-term Downward Trend in Total Solar Irradiance, *Science*, **234**, 1114.

Woodard, M. 1984: Short-Period Oscillations in the Total Solar Irradiance, *Ph.D. Thesis, Univ. Calif. at San Diego*, La Jolla, CA.

CAUSES OF INTERANNUAL GLOBAL TEMPERATURE VARIATIONS OVER THE PERIOD SINCE 1861

P.D. Jones and P.M. Kelly
Climatic Research Unit
School of Environmental Sciences
University of East Anglia
Norwich NR4 7TJ
United Kingdom

Hemispheric and Global Temperature Data

Understanding of the past record of global surface air temperature variations has been improved significantly by two recent
developments. First, the land-based record has been extended
back in time and into areas where data were previously sparse
and the station data on which it is based have been tested for
homogeneity and corrected or rejected when found to be unsatis-
factory. Second, data from ocean areas have been incorporated,
markedly improving the spatial representativeness of the global
record.

As far as the land-based record is concerned, there has been
an overall increase in the number of stations used. All of this
new station temperature data has been unearthed through ex-
haustive searches of meteorological archives, particularly those
of the National Meteorological Library in Bracknell, U.K. The
individual temperature series have been assessed, where possible, for homogeneity. Many station records contain fluctuations that result from non-climatic factors such as station
moves, changes in instrumentation, and so on (Jones et al.,
1986a). Some station records required correction and others had
to be omitted because of these problems. Complete details of all
the stations used, the homogeneity analyses and the corrections
applied are available in a series of technical reports (Bradley
et al., 1985; Jones et al., 1985, 1986b).

Data from usable stations, expressed as departures from the
1951-70 reference period mean, were interpolated onto a regular
latitude-longitude grid in order to overcome the irregular station network. Figure 1 shows annual mean temperature estimates
for the Northern and Southern Hemispheres since the middle of
the last century. Note that the Southern Hemisphere curve only
includes data for the Antarctic since 1957.

Meteorological observations from the land areas of the hemispheres alone may not give a true picture of hemispheric temperature. Only about 30% of the area of the globe is land. For
the marine areas, it is necessary to use observations taken by
so-called 'ships of opportunity'. The most complete compilation
of historical marine instrumental data is known as COADS (Comprehensive Ocean Atmosphere Data Set) (Slutz et al., 1985;
Woodruff, 1986). This data set contains all types of marine observations, the most numerous of which are the approximately
63.25 million observations of sea surface temperature (SST). It
covers the years 1854 to 1979.

Marine observations are, like the land data, subject to homogeneity problems (Barnett, 1984; Goodess and Kelly, 1987). In
the case of the marine data, the problems are, by their very
nature, more difficult to overcome than for the land-based
series. Changes in instrumentation, in the way in which observations at sea are taken and in the size, speed and fabric of
ships have all occurred affecting the homogeneity of the marine
database. In many cases, information about the method of observing has been lost or was never recorded. The most well known
problem is the change in the method of measuring SST from the
use of a thermometer in a bucket of sea water to the use of a
thermometer located in the intake pipes supplying water for
cooling to the ship's engines. Readings using the latter
measurement technique have been shown to be between 0.3 and 0.7°C
warmer than the earlier bucket method. For most SST observations
in the data base it is not known how each observation was made.
It is generally assumed that bucket measurement prevailed before
1940 and intake measurement was more common after that time
(Barnett, 1984).

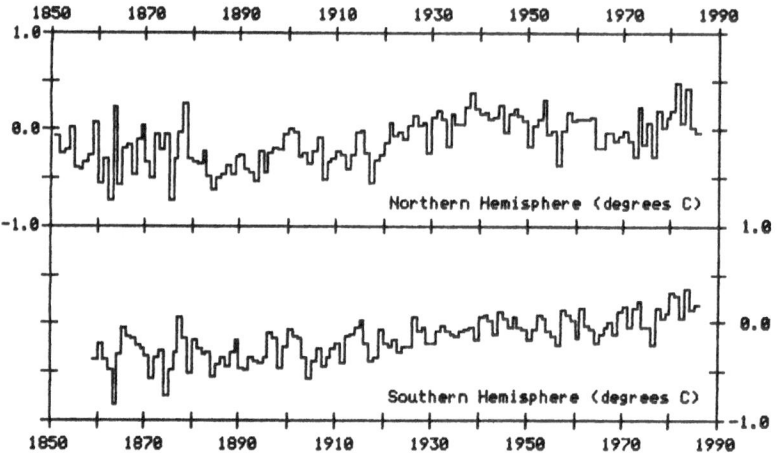

Fig. 1: Annual temperature estimates from land-based data for
        the Northern and Southern Hemispheres. Data are ex-
        pressed as anomalies (degrees Celsius) from the 1951-
        1970 reference period (see Jones et al., 1986a,c).

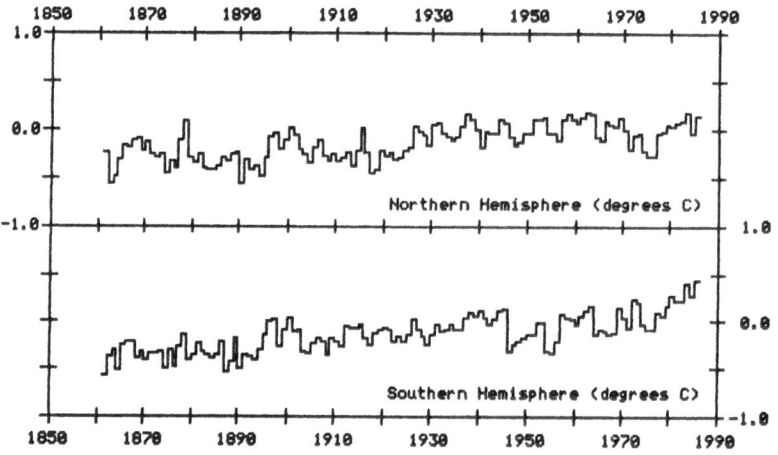

Fig. 2: Annual temperature estimates from SST observations for
        the Northern and Southern Hemispheres. Data are ex-
        pressed as anomalies (degrees Celsius) from the 1950-
        1979 reference period (see Jones et al., 1986d). Up-
        dates of the SST series for the final six years have
        been   made   from   adjusted   Climate   Analysis   Center
        analyses (Reynolds and Gemmill, 1984).

Large-scale averages of marine air temperature (MAT) measured
on board ships have been homogenised by Jones et al. (1986d).
Hemispheric and regional temperatures based on COADS have been
compared with those based on the more reliable land-based data.
Intuitively, hemispheric estimates from the two data sets should
be the same (Wigley et al. 1985). Thus, any systematic differences between the series may be used to correct the MAT series.
The consistency of the differences between the two estimates
through time and between hemispheres justifies this approach.
Once the MAT series has been corrected, a similar comparison
technique may be used to correct the SST series. Agreement
between hemispheric SST and MAT estimates should be even better
than in the case of the MAT/land comparison.

Figure 2 shows annual mean SST anomalies for the Northern and
Southern Hemisphere since 1854. The hemispheric MAT variations
are very similar to those of SST (r^2 over 1861-1979 for NH and
SH: $\sim$ 0.85). Whilst the curve for the Northern Hemisphere is
representative of all northern oceans except the Arctic Ocean,
the Southern Hemisphere curve is only representative of the
southern oceans between the equator and 45°S. There are practi-
cally no data available between 45°S and Antarctica except near
the southern tip of South America.

Combining the land and ocean data is relatively straight-
forward. Figure 3 shows combined land and ocean (SST) series for
the Northern and Southern Hemispheres. The homogenised hemispheric series for MAT or SST may be used to represent the ocean
areas. For the Northern Hemisphere, land and ocean portions are
weighted equally while for the Southern Hemisphere the ocean is
weighted 1.5 times the land in order to account for the different area of land and ocean. Figures 1, 2 and 3 show many consistent features, in particular, the similarities between the
Northern and Southern Hemispheres. The warming trend exhibited
by these series is consistent with model predictions of the
effects of increasing atmospheric levels of greenhouse gases
(Wigley et al., 1986).

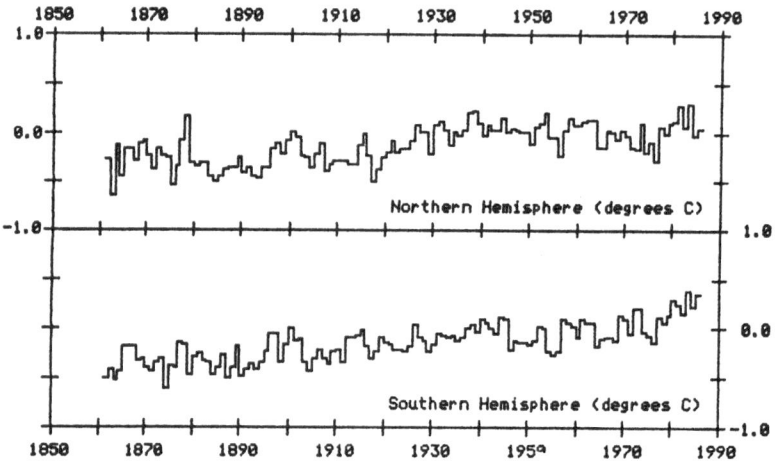

Fig. 3: Annual temperature estimates from a combination of land-based data and SST observations for the Northern and Southern Hemispheres. Data are expressed as anomalies (degrees Celsius) from the 1950-1979 reference period (see Jones et al., 1986d).

Possible forcing factors

Ever since the first estimates of global mean temperature were made (Köppen,1873), causal mechanisms have been proposed to explain the observed variations on year-to-year and longer time-scales. The two most commonly considered mechanisms or forcing factors have been variations in solar activity and volcanic activity (see, for example, Köppen, 1914). These factors, together with the El Niño/Southern Oscillation (ENSO) phenomenon and potential effects of increasing carbon dioxide, are considered to be the most probable causes of variations in global mean temperature on the 1 to 100 year time scale (see Wigley et al., 1985, for a recent review). Here, we consider and compare the effects of two of these factors, volcanoes and the ENSO phenomenon, on hemispheric temperature.

Volcanic effects

Explosive volcanoes can inject ash and gas into the upper atmosphere. Once ash reaches the stratosphere it can remain for up
to two years. Over the initial months, sulphur gases are trans-
formed into sulphate aerosols in a process known as secondary
aerosol formation. The ash and sulphate aerosols scatter incoming energy from the sun and thus reduce the amount of solar
radiation that reaches the surface of the Earth, by one or two
percent in the case of a large eruption. The net effect should
be to cool the surface whilst the volcanic aerosols are resident
in the stratosphere (Lamb, 1970). Volcanoes which only inject
material into the troposphere, where it is readily washed out by
precipitation, are only likely to affect the weather in the
immediate vicinity of the volcano for a few days.

Table 1: Selected Volcanic Events

Event	Year	Month	Latitude	Longitude	VEI[*]
Pelee/Soufriere	1902	5	~ 14 N	61.2W	4
Ksudach	1907	3	51.8N	157.5E	5
Novarupta	1912	6	58.3N	155.2W	6
Bezymianni	1956	3	57.1N	160.7E	5
Krakatau	1883	8	6.1S	105.4E	6
Tarawera	1886	6	38.2S	176.5E	5
Azul	1932	4	35.7S	70.8W	5
Agung	1963	3	8.3S	115.5E	4

*Volcanic Explosivity Index

NOTE: Two volcanic eruptions in the Caribbean (Pelee and
Soufriere) occurred during May 1902. Both eruptions were VEI =
4. The eruption later that year in Guatemala (VEI = 5) was not
used in this analysis because the effects are likely to be ob-
scured by the earlier eruptions in that year. The Agung eruption
was included because of its high Dust Veil Index (Lamb, 1970).
See Kelly and Sear (1984) for further details.

In order to determine the volcanic influence on climate, it is necessary to assess the magnitude of past volcanic eruptions. Various workers have compiled eruption catalogues on the basis of historical accounts, geological evidence, and so on (e.g. Lamb, 1970; Simkin et al. 1981). Inevitably, assessment of eruption size and potential climate effect is fraught with uncertainty because of the limited information that is availabe. Nevertheless, it is possible to identify the historical eruptions most likely to have affected climate. On the basis of a careful synthesis of published material, Kelly and Sear (1984) identified eight volcanoes during the period 1881-1980 which were likely to have had significant climatic effects. The selection drew heavily on the geological catalogue of Simkin et al. (1981) and used supplementary information from Lamb (1970). Table 1 lists these volcanoes and the date of the major eruption.

To examine the effect of these eruptions on hemispheric mean temperature, a superposed epoch analysis (Conrad and Pollak, 1962; see also Kelly and Sear, 1984, and Sear et al., 1987) was carried out. First, hemispheric temperature estimates for the 60 months after each eruption date were expressed, month by month, as departures from the mean hemispheric temperature for the 36 months before the event, the prevailing temperature level. A composite response was then formed by averaging the four individual responses. Averaging in this way emphasises the common features. The results are shown in Figure 4a (Northern Hemisphere eruptions) and Figure 4b (Southern Hemisphere eruptions). We have analysed the combined land and ocean temperature series for the Northern and Southern Hemisphere. A Monte Carlo approach was used in order to assess the statistical significance of the results. 500 analyses of four randomly chosen events between 1861 and 1980 were used to determine 5% significance levels. One-tailed tests have been used as volcanoes are unlikely to cause surface warming. The significance level is plotted as a horizontal dashed line.

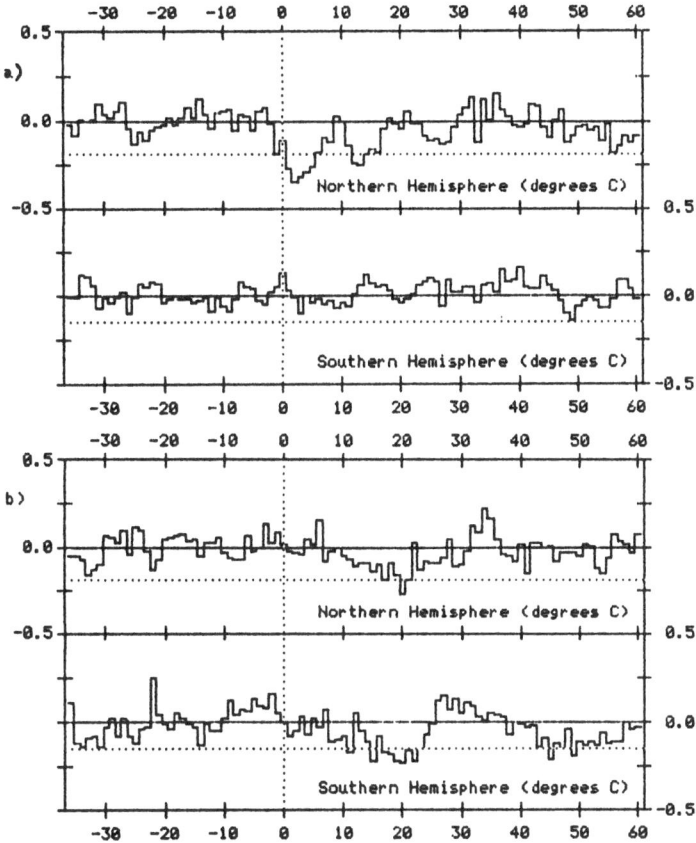

Figure 4: Superposed epoch analysis of monthly hemispheric temperature data (land and SST estimate). (a) Northern Hemisphere eruptions; (b) Southern Hemisphere eruptions.

The effect of Northern Hemisphere eruptions on Northern Hemisphere temperature is extremely rapid - the maximum response of about 0.3 °C occurs within a few months of the eruption. The slight cooling prior to month zero can be considered noise. The effect of Southern Hemisphere eruptions on southern temperatures is delayed with the maximum effects, about 0.2°C. Northern eruptions appear to have a negligible effect on temperatures in the Southern Hemisphere but southern eruptions cool the Northern Hemisphere significantly, by about 0.15°C, some two years after the event.

Whether these differences in response result from the location of this particular sample of eruptions, their seasonal distribution, or from the relative percentages of land and ocean in each hemisphere is not certain. The most likely explanation is that the more rapid response in the Northern Hemisphere is due to the greater land area. The land is able to respond more rapidly than ocean areas owing to their lower thermal inertia. The oceans of the Southern Hemisphere tend to delay or reduce the magnitude of the volcanic effect. However, most of the Northern Hemisphere eruptions were in higher latitudes and the pollutants were less likely to have been transported to the Southern Hemisphere in large quantities. Most of the Southern Hemisphere eruptions were in the southern equatorial zone and substantial transport of material to the Northern Hemisphere is known to have occurred. Remarkable sunsets were seen in Europa after the Krakatau eruption (Lamb, 1970). Finally, most of the eruptions were in the northern spring or summer and this may also have affected the characteristics of the response. Bradley (1987) has noted that the volcanic effect on northern temperatures is seasonally-dependent. This can be seen in the upper curve in Figure 4a.

The El Niño/Southern Oscillation Phenomenon

El Niño events in the tropical Pacific Ocean are now known to be of global significance (Bjerknes, 1969; Rasmussen and Carpenter, 1982). These warming events are associated with one extreme of the Southern Oscillation (Walker and Bliss, 1932; WMO, 1985), characterised by negative departures in the relevant indices. The Southern Oscillation is a measure of major shifts in the pressure distribution of the central and southern Pacific, the so-called Walker Circulation. The complementary positive extremes of the Southern Oscillation (cold events) may also be of global significance (Berlage, 1957). Taken together, these variations are collectively referred to as the El Niño/Southern Oscillation (ENSO) phenomenon.

The most commonly used index of the Southern Oscillation is the difference in monthly mean sea level pressure between

Tahiti, Society Islands, and Darwin, Australia. This Southern
Oscillation Index (SOI) has recently been extended to 1882 by
Ropelewski and Jones (1987). Here, we use a further extension
back to 1866 based on data from Djakarta, Indonesia, in place of
Darwin and Santiago, Chile, in place of Tahiti. Annual values of
this index are plotted in Figure 5.

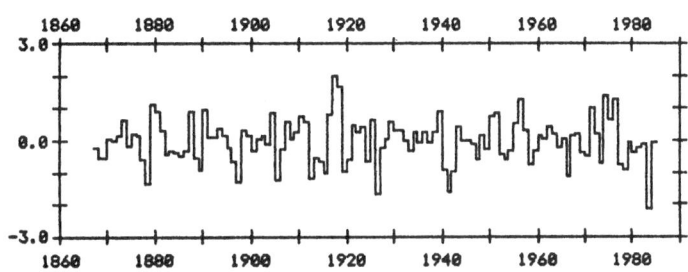

Figure 5: Annual (July-June, dated by the January) values of the
extended Southern Oscillation Index.

El Niño is best defined by the occurrence of extremely warm or
extremely cold water off the coast of Peru and Ecuador in South
America. Using the Quinn et al. (1978) catalogue of El Niño
Peruvian coastal SST variation and the SOI, Bradley et al.
(1987) have identified 23 warm events (El Niños) and 20 cold
events between 1880 and 1980. The years are listed in Table 2.

The relationship between the ENSO indicators and hemispheric
mean temperature is examined in two ways: first, by direct correlation and, second, through the use of superposed epoch analysis.

Table 2: <u>ENSO Warm and Cold Events</u>

Warm (El Niño) years:

1884	1888	1891	1896	1899
1902	1904	1911	1913	1918
1923	1925	1930	1932	1939
1951	1953	1957	1963	1965
1969	1972	1976		

Cold years:

1886	1889	1892	1898	1903
1906	1908	1916	1920	1924
1928	1931	1938	1942	1949
1954	1964	1970	1973	1975

a) Direct Correlation

Coefficients of determination were computed between the extended SOI and 12-month hemispheric temperature averages with the temperature series lagging the SOI by monthly intervals up to 12 months. The results are shown in Figure 6. The analysis was performed over two periods, 1867-1925 and 1926-1984, to check the stability of the relationships. Slightly more variance was explained by the relationship during the earlier period.

The relationship is strongest when the SOI leads the temperature series by about six months. Between 25 and 30% of the variance of the hemispheric temperature series is then accounted for by the SOI. This lag relationship of about six months may be of some forecasting value. Global temperatures should be lowered/raised by 0.05°C for each +ve/-ve departure of one unit of this particular SOI. The 1982/3 event, with an SOI of ~ -3, should have increased global temperatures by 0.15°C, enough to mask any immediate effect of the eruption of El Chichon in spring 1982.

b) Superposed epoch method

Superposed epoch analysis was applied in the same way as in the volcanic case. The Januarys of the warm (El Niño) and cold events listed in Table 2 were used as the key months in two separate analyses. The results are shown in Figures 7a (warm)

and 7b (cold). Two-tailed significance levels were assessed
using the Monte Carlo technique by randomly selecting the appro-
priate number of events. The significance levels are much lower
here, compared to the volcanic case, because there are more
events.

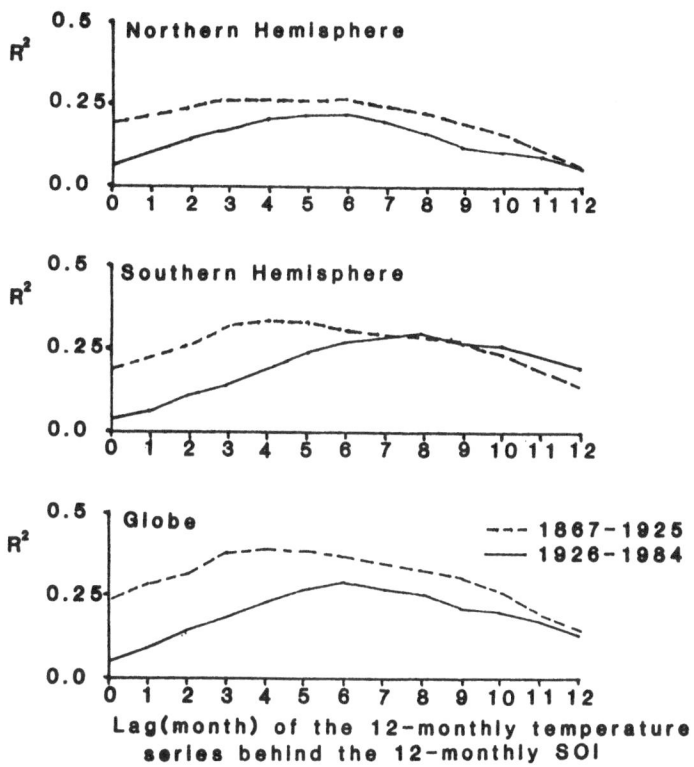

Figure 6: Lagged coefficients of determination between the
Southern Oscillation Index and hemispheric
temperature series (land and SST estimate).

Most warm (El Niño) events tend to commence around the turn
of the year with the SOI used here showing the greatest response
during the following June or July. The timing is similar for the
cold periods. Figure 7 shows that the maximum effect on hemispheric temperature occurs some 10 to 16 months after the January of the selected year. The results are compatible with those
of Figure 6 when one considers that the temperature response
lags six months behind the SOI which, in turn, lags the January
of the year of the warm or cold event by some six months. The
effect on temperature of the warm and cold events is similar: a

warming/cooling of 0.1 to 0.2 °C some 10 to 16 months after the
January key date. Note the apparent precursor of a warm event:
low temperatures during the previous year (Figure 7a, see also
Bradley et al., 1987).

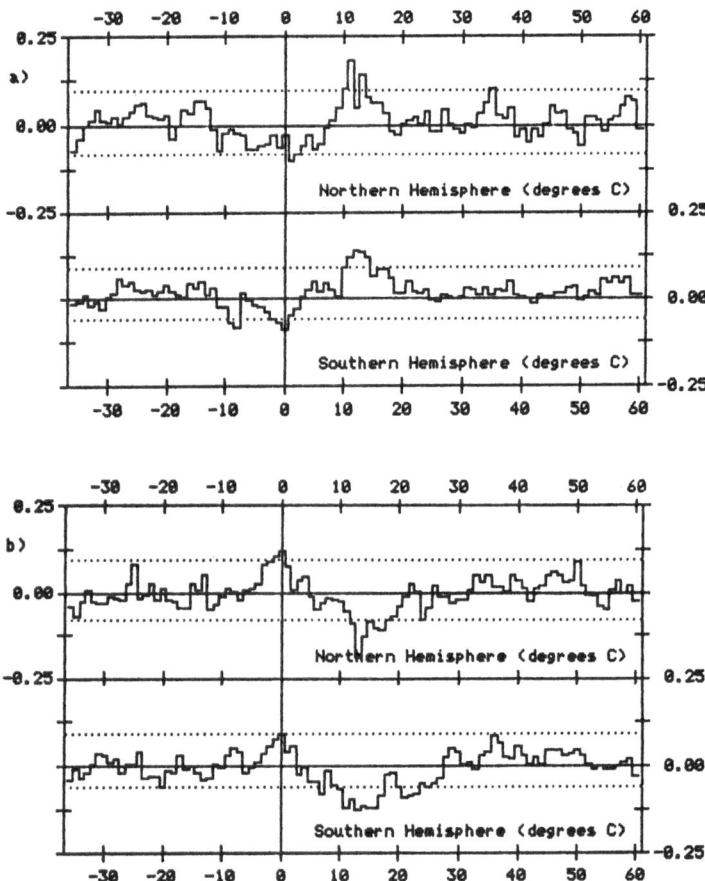

Figure 7: Superposed epoch analysis of monthly hemispheric
           temperature data (land and SST estimate). (a) Warm
           (El Niño) events; (b) cold events.

Conclusions

Large explosive volcanic eruptions and the ENSO phenomenon have been shown to have similar effects, of the order of 0.1 to 0.2$\overset{\circ}{\text{C}}$, on hemispheric temperature. The effects are relatively short-lived. The duration of the maximum effect is of the order of 6 months and it occurs at some time within the two years immediately after the volcano or warm or cold event. These two factors are responsible for between 30 and 50% of the inter-annual, high frequency variability in the hemispheric temperature records.

Acknowledgements

This work was supported by the United States Department of Energy under contract number DE-FG02-85ER60316. The authors acknowledge the comments of Ms. C.M. Goodess on an earlier draft of the manuscript.

References

Barnett, T.P., 1984: Long term trends in surface temperature over the ocean. Monthly Weather Review 112, 303-312.

Berlage, H.P., 1957: Fluctuations in the general atmospheric circulation of more than one year; their nature and prognostic value. KNMI Meded. Verh 69.

Bradley, R.S., 1987: The explosive volcanic eruption record in Northern Hemisphere continental temperature records. Climatic Change (submitted).

Bradley, R.S., Kelly, P.M., Jones, P.D., Diaz, H.F. and Goodess, C., 1985: A climatic data bank for the Northern Hemisphere, 1851-1980, DoE Tech. Rep. No. TR017, Carbon Dioxide Research Division, Washington, D.C. 335 pp.

Bradley, R.S., Diaz, H.F., Kiladis, G. and Eischeid, J.K., 1987: ENSO Signal in continental temperature and precipitation records. Nature (submitted).

Conrad, V. and Pollak, L.D., 1962: Methods in Climatology, Harvard University Press, 459 pp.

Goodess, C. and Kelly, P.M., 1987: Historical marine air temperature and sea surface temperature data. A review of the data quality problem and implications for the calculation of global averages. DoE Tech. Rep., U.S. Dept. of Energy, Carbon Dioxide Research Division.

Jones, P.D., Raper, S.C.B., Santer, B.D., Cherry, B.S.G., Goodess, C., Bradley, R.S., Diaz, H.F., Kelly, P.M. and Wigley, T.M.L., 1985: A grid point surface air temperature data set for the Northern Hemisphere, 1851-1984. DoE Tech. Rep. TR022. U.S. Dept. of Energy, Carbon Dioxide Research Division, Washington, D.C. 251 pp.

Jones, P.D., Raper, S.C.B., Bradley, R.S., Diaz, H.F., Kelly, P.M. and Wigley, T.M.L., 1986a: Northern Hemisphere surface air temperatures: 1851-1984. Journal of Climate and Applied Meteorology, 25, 161-179.

Jones, P.D., Raper, S.C.B., Cherry, B.S.G., Goodess, C. and Wigley, T.M.L., 1986b: A grid point surface air temperature data set for the Southern Hemisphere, 1851-1984. DoE Tech. Rep. TR027. U.S. Dept. of Energy, Carbon Dioxide Research Division, Washington, D.C. 73 pp.

Jones, P.D., Raper, S.C.B. and Wigley, T.M.L., 1986c: Southern Hemisphere surface air temperatures: 1851-1984. Journal of Climate and Applied Meteorology, 25, 1213-1230.

Jones, P.D., Wigley, T.M.L. and Wright, P.B., 1986d: Global temperature variations between 1861 and 1984. Nature, 322, 430-434.

Kelly, P.M. and Sear, C.B., 1984: Climatic impact of explosive volcanic eruptions. Nature, 311, 740-743.

Köppen, W., 1873: Über mehrjährige Perioden der Witterung, insbesondere über die 11-jährige Periode der Temperatur. Zeitschrift der Österreichischen Gesellschaft für Meteorologie, 8, 241-248; 257-267.

Köppen, W., 1914: Lufttemperaturen, Sonnenflecken und Vulkanausbrüche. Meteorologische Zeitschrift, 31, 305-328.

Lamb, H.H., 1970: Volcanic dust in the atmosphere, with a chronology and assessment of its meteorological significance. Phil. Trans. of the Royal Society of London, 266A, 425-533.

Quinn, W.H., Zopf, D.O., Short, K.S. and Kuo Yang, R.T.W., 1978: Historical trends and statistics of the Southern Oscillation, El Niño and Indonesian droughts. Fisheries Bulletin, 76, 663-678.

Rasmussen, E.M. and Carpenter, T.H., 1982: Variations in tropical sea surface temperature and surface wind fields associated with the Southern Oscillation/El Niño. Monthly Weather Review, 110, 354-384.

Reynolds, R.W. and Gemmill, W.H., 1984: An objective global monthly mean sea surface temperature analysis. Tropical Ocean-Atmosphere Newsletter, 23, 4-5.

Ropelewski, C.F. and Jones, P.D., 1987: An extension of the Southern Oscillation Index. Monthly Weather Review (in press).

Sear, C.B., Kelly, P.M., Jones, P.D. and Goodess, C.M., 1987: On the response of global surface air temperatures to major volcanic eruptions. Nature, (submitted).

Simkin, T. et al., 1981: Volcanoes of the World. Hutchinson Ross, Stroudsberg.

Slutz, R.J., Lubker, S.J., Hiscox, J.D., Woodruff, S.D., Jenne, R.L., Joseph, D.H., Steurer, P.M. and Elms, J.D., 1985: Comprehensive Ocean Atmosphere Data Set Release 1 NOAA, Environmental Research Laboratories, Boulder, Colorado. 268 pp.

Walker, G.T. and Bliss, E.W., 1932: World Weather V Memoirs of the Royal Meteorological Society IV (36), 53-84.

Wigley, T.M.L., Angell, J.K. and Jones, P.D., 1985: Analysis of the temperature record. In: M.C. MacCracken and F.M. Luther (Eds.), Detecting the effect of increasing atmospheric CO_2. U.S. Dept. of Energy Carbon Dioxide Research Division, pp. 55-90.

Wigley, T.M.L., Jones, P.D. and Kelly, P.M., 1986: Empirical Climate studies: warm world scenarios and the detection of climatic change induced by radiatively active gases. In: B. Bolin, J. Jäger, B.R. Döös and R.A. Warrick (Eds.). The Greenhouse Effect, Climatic Change and Ecosystems. SCOPE 29, Wiley, pp. 271-322.

World Meteorological Organization, 1985: The Global Climate System: A critical review of the climate system during 1982-84. World Climate Data Programme.

Woodruff, S.D., 1986: Editor, Proceedings of a COADS workshop, Boulder, Colorado, January 22-24, 1986. NOAA Technical Memorandum ERL ESG-23, 218 pp.

CLIMATIC INFORMATION FOR THE PAST HUNDRED YEARS IN WIDTH AND DENSITY OF CONIFER GROWTH RINGS

F.H. Schweingruber
Swiss Federal Institute of Forestry Research
CH-8903 Birmensdorf

1. Introduction

Growth rings of trees growing in areas with a seasonal climate contain climatological information. There is an extensive literature on this subject, e.g. Fritts (1976). However, although many studies have been made, most dendroclimatological investigations in the non-arid zones of the earth have met with little success and reconstructions on a year-by-year basis have seldom proved possible. All reconstructions so far attempted have been formulated in terms of one or another moving average over a defined period, e.g. a decade, and have almost exclusively been based on measurements of ring width. Those limitations have recently given rise to new approaches, which are discussed below.

- Greater attention is being paid to the selection of sites and individual trees, since it has been shown that site factors exert a stronger influence on growth ring formation than regional climatic conditions, especially in areas with a temperate climate.

- To allow due consideration of the great variation in growth ring anatomy, the long-established procedure of dating through pointer years has been brought into relationship to climatic and ecological factors, with proper attention to abrupt changes in ring width; radiodensitometric methods have been expanded; tissue analysis has been refined; and isotope research related to growth rings and climate has been undertaken.

Figures la und lb show the morphological features used as criteria in dendroclimatological research.

The examples given below illustrate how these new methods can greatly extend the present knowledge on dendroclimatology and that growth rings, as sources of proxy data, can supply information on climatic conditions over long periods of time and great geographical distances.

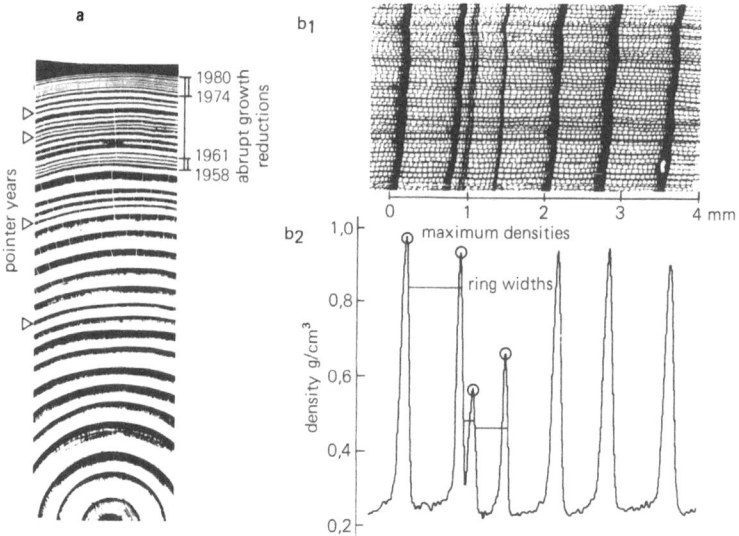

Fig. 1a

Growth ring sequence from a fir displaying pointer years (growth changes within one or two years) and abrupt growth reductions (persisting for several years). Those changes which can be identified and dated by the naked eye are expressions of severe changes in the physiology of the tree. Such changes are often triggered by extreme effects of the climate (summer drought, cold periods, extremely low temperatures in winter, etc.).

Fig. 1b

Parallel diagrams from a larch sequence with pointer years; b1: photomicrograph, b2: corresponding density profile. Five different parameters of such curves are selected for interpretation. Maximum density and ring width contain by far the greatest climatological information.

2. Temperature reconstructions from conifers in Europe

Studies by Parker and Hennoch (1971), Schweingruber et al. (1978) and Hughes et al. (1984) have shown that the maximum density of different species of conifer growing on cold and wet sites in the Alps and in Scotland generally contain information on temperature during the summer months from July to September. Response functions for maximum density chronologies from northern Scandinavia provide data on temperature during July and August, while those from the mountain ranges of Central and Southern Europe supply information on conditions during July, August and September. Ring width, on the other hand, is less clearly related to weather conditions, since it is far more strongly influenced by local site conditions prevailing during the relevant vegetation period and the preceding year. Maximum density integrates climatological information to a greater degree than ring width, because it is mainly an expression of the cell wall thickness of the latewood cells. Since these survive for 2 to 3 months, even at the upper and northern timberlines, they are able to reflect limiting growth factors, in this case temperature. Ring width, in contrast, is an expression of the performance of the cambium, whose main activity is limited to two to four weeks in early summer.

These considerations and observations led to the construction of an ecologically uniform sample net. In order to maximize the reflection of temperature in the growth-ring patterns, samples were taken from normally grown trees on the coldest and the wettest sites in both the upper and the northern timberline zones. 101 local indexed chronologies (2 cores from each of 12 trees per site) were used to construct 22 regional chronologies. The growth-ring data (maximum density) were calibrated by comparing maps of annual anomaly patterns of densities or their index values, with maps of the corresponding temperature anomalies (= departures form the the long-term average). The reference period for maximum density was the overall temporal range of the chronologies, that for summer temperature the reference period 1881-1980 considered by Jones et al. (1982).

Visual comparison of the maps revealed similar to very similar patterns in 4/5 of the cases (Fig. 2, Appendix). Discrepancies can be explained through gaps in the meteorological data and through differences in growth limiting factors in various regions of Europe, for instance, the growth period is limited to July and August in northern

Scandinavia but extends from July to October in southern Italy. Furthermore, the possibility cannot be excluded that in years with extreme conditions, e.g. 1948, growth was not limited by temperature
alone on some sites.

Similar growth maps were constructed on the basis of ring width,
but it has not yet proved possible to decode the climatological information they contain in year-by-year terms.

This study shows that density analysis permits very reliable
reconstruction of temperature patterns for the whole Northern Hemisphere (boreal zone, upper timberline in mountains). Reconstruction of
precipitation over fairly large areas is equally possible. The con-
dition, however, is that the samples be uniformly taken the lower
timberline zone, where precipitation is essentially the only growth-
limiting factor (Fig. 3; Fritts 1974, Stockton et al. 1981). Networks
comprising trees on differing sites are unlikely to provide uniform
climatological signals. Consequently, year-by-year reconstructions
cannot be made on the basis of data from such networks, and reconstructions for longer periods are always accompanied by broad, often
incalculable deviations (Shiyatov and Mazepa 1986, Fritts 1974).

3. Limitations of dendroclimatological research

The influence of growth-limiting factors varies greatly between species and between sites, especially on non-extreme sites. In order to
study such relationships, Lingg (1986) investigated the differences
between spruce and fir over the past 80 years along altitudinal tran-
sects in the Valais (Fig. 4). Ecophysiological behaviour differs quite
considerably from species to species. In firs both maximum density and
ring width in trees growing on low and high altitude sites correlate
with each other. Only in trees of the subalpine zone does maximum
density fail to reflect differences between site or species. It seems
probable that too little attention is being paid to the difference in
ecophysiological behaviour between species in the construction of
dendroclimatological networks.

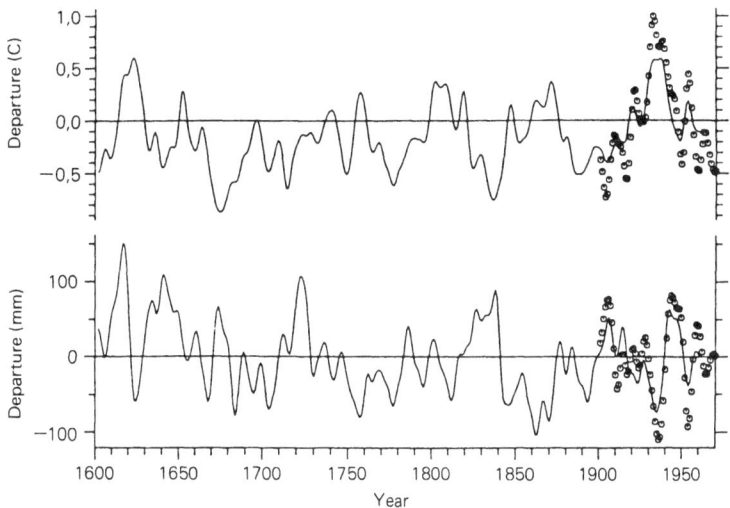

Fig. 3

Reconstructed fluctuations in temperature and precipitation based on
new ring width chronologies from the Great Plains, USA. Relationships
between ring widths and recorded meteorological parameters are calibrated for the period 1900-1970. (Circles: meteorological measurements, lines: reconstructions, smoothed with a low pass filter). After
Fritts (1983).

    Climatic differences, varying from year to year, are strongly
modified by local site factors and affect ring formation accordingly.
In a radiodensitometrical study Kienast 1985 clearly shows the relationship between different growth ring parameters and relief.

    The influence of site factors on growth ring formation in the
mountainous areas of the temperate zone is summarized below (Fig. 5).

- Where the regional climate is cold and moist, growth on moist, high
  altitude sites is severely limited, while growth on shallow-soil low
  altitude sites is optimum (Type A).

- Where regional climate is warm and dry the situation is reversed:
  growth is optimum on moist, high altitude sites but minimal on shallow-soil low altitude sites (Type B).

- Where regional climate is cold and dry, growth is optimum on both
  dry, high altitude sites and moist, low sites, but limited on moist,
  high altitude sites by the low summer temperatures and on shallow,
  low altitude sites by low precipitation (Type C).

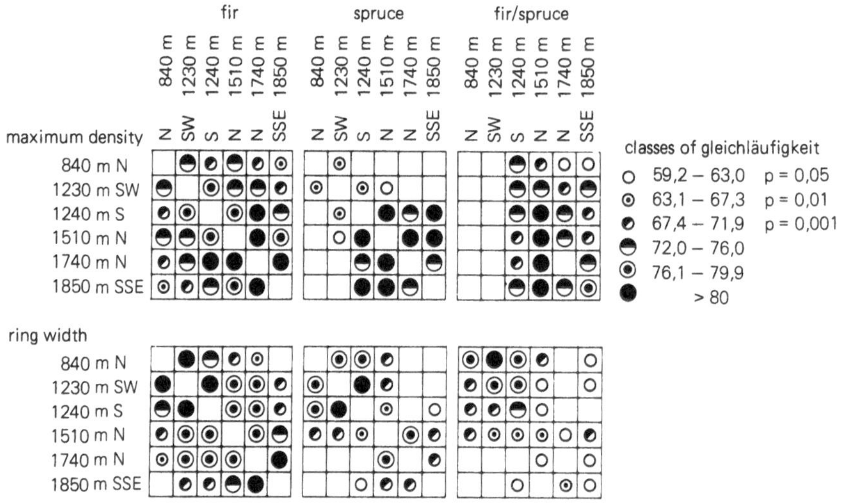

Fig. 4

Relationships (Gleichläufigkeit) of maximum density and ring width between firs (Abies alba; left), spruces (Picea abies; center), and (right) between firs (vertical) and spruce (horizontal), along an altitudinal profile in the Valais, Switzerland for the period 1900–1980. Firs and spruces are from common sites; cores were taken from 12 firs and 12 spruces at each site. After Lingg (1986).

The difference in behaviour between the two species and the ring parameters is evident. Firs display greater similarity over a fairly wide spectrum than spruces. This may be mainly due to the different types of root systems. While the deep root system of fir allows an efficient water supply throughout the year, the shallower, more super-ficial one of spruce often leads to growth inhibition through low pre-cipitation, particularly on low altitude sites.

Fir: maximum density: trees on all sites behave similarly.
         ring width   : trees on sites above 840 m behave similarly,
                          the low altitude site (840 m) behaves differ-
                          ently.

Spruce: maximum density: the trees on the lowest two sites (840 and
                          1230 m), the two sites at 1230 m and 1240 m,
                          and the three highest sites display relation-
                          ships among themselves.
         ring width   : the trees of the four lowest (840–1510 m) and
                          two highest sites display relationships among
                          themselves.

Fir/spruce relationship
         maximum density: the two lowest sites differ clearly from all
                          the higher ones.
         ring width   : the pattern is similar to that of spruce
                          alone.

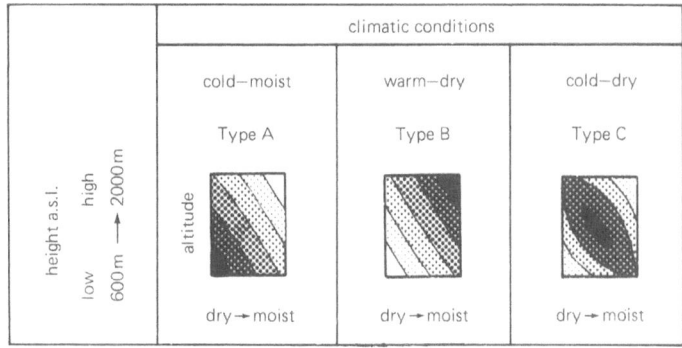

Fig. 5

Idealized dendro-ecological diagram, showing the relationships between different altitudes (ordinates) and site moisture (abscissae) during years with differing weather patterns. Optimum ring growth is represented by black shading, minimum growth by white. After Kienast (1985).

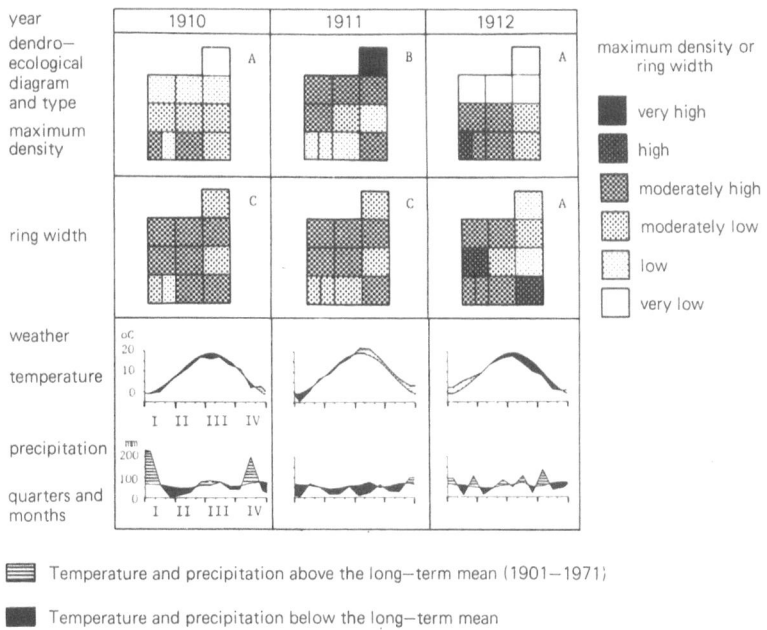

Fig. 6

Temporal series of dendro-ecological diagrams for maximum density and ring width in trees growing in the Valais (Switzerland) and the nearby meteorological station at Sitten. Arrangement of sites (squares) as in Figure 5 (vertical: altitude, horizontal: dry-moist). Growth at high altitude sites was considerably reduced during the cold years 1910 and 1912, while the dry conditions of 1911 impaired growth on the lower sites. After Kienast (1985).

Fig. 6, however, shows that the limiting factors vary from year
to year in their effect on the specific ring parameter according to
the weather pattern. That means that it is possible to extract information on different climatic components from one and the same growth
ring sequence, provided that a number of site chronologies are considered in relation to each other. Unfortunately, the procedure normally
employed in dendrochronology for the analysis of time series (response
functions) does not permit the limiting ecological factors to be deco-
ded on a year-by-year basis.

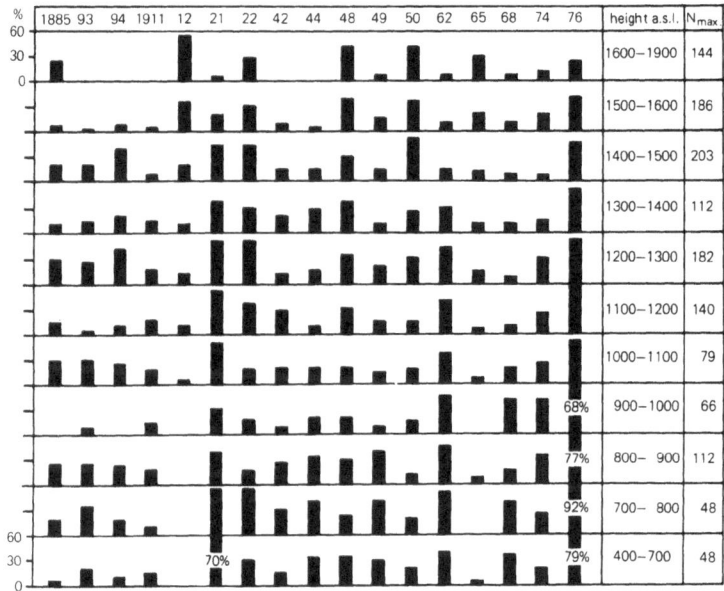

Fig. 7

Altitudinal distribution of pointer years in spruce in the Valais,
Switzerland, since 1885. The height of the columns represents the
percentage of spruces at particular altitudes with pointer years
($N_{max.}$: maximum number of spruces investigated). It is noticeable
that pointer years are only formed on high altitude sites during cold
years, e.g., 1912. In non-extreme dry years they are mainly formed on
sites at lower altitudes, e.g., 1942, 1944. After Kontic et al.
(1986).

Altitudinal relationships have been described by Kontic et al. (1986), the interpretation being mainly based on pointer years, that is, growth rings which, for the majority of trees on a phytosociologically homogeneous site, are markedly wider or narrower than the preceding or subsequent ones (Fig. 7).

In the upper timberline zone, tree growth is mainly limited by low temperature during the vegetation period, as happened in 1912 and 1965, for example. At lower elevations the limiting factor is precipitation during summer, witness 1921, 1942, 1944, 1949, 1976 etc. On medium altitude sites in the temperate zone the limiting factors vary greatly, indeed to such an extent that it is not yet even possible to explain the occurrence of pointer years coinciding over wide areas.

These three cases show the overriding necessity of extreme care in selecting sites for dendroecological studies.

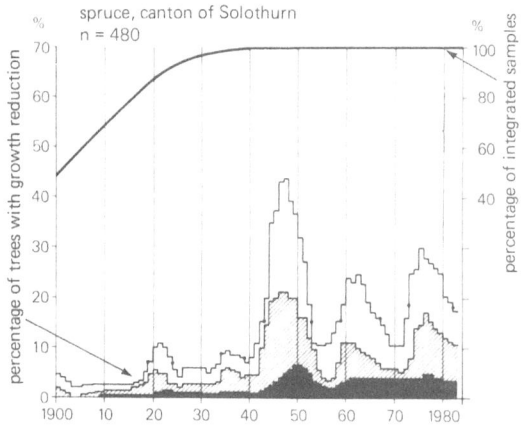

Fig. 8

Summation diagram for spruces with growth reduction in the canton of Solothurn, Switzerland. The diagram shows the percentage of trees whose growth has been reduced since 1910 in comparison to the preceding period (black: over 71%, hatched: 56-70%, white: 40-55% growth reduction). The fluctuations are evident. The phase of growth reduction between 1945 and 1954 is conspicuous.

Abrupt growth changes were long regarded in dendrochronology as disturbances, since they often reflect individual changes such as disease, injury to the photosynthetically active crown, or alteration in the vertical position of the tree. Recent studies, however, have shown that abrupt growth changes persisting for more than three years incorporate climatic signals.

A supra-regional study of growth patterns over the present century in several thousand conifers of different species in Switzerland revealed a certain trend towards a periodicity of 11-16 years, which seems to be mainly governed by deficits in summer precipitation (Figs. 8, 9). It is quite obvious that some of these growth changes are due to local pollution, disease, or impairment of the site through soil compaction or sinking of the ground water level. It is a task for dendrochronological research to clarify the origin of these irregularities.

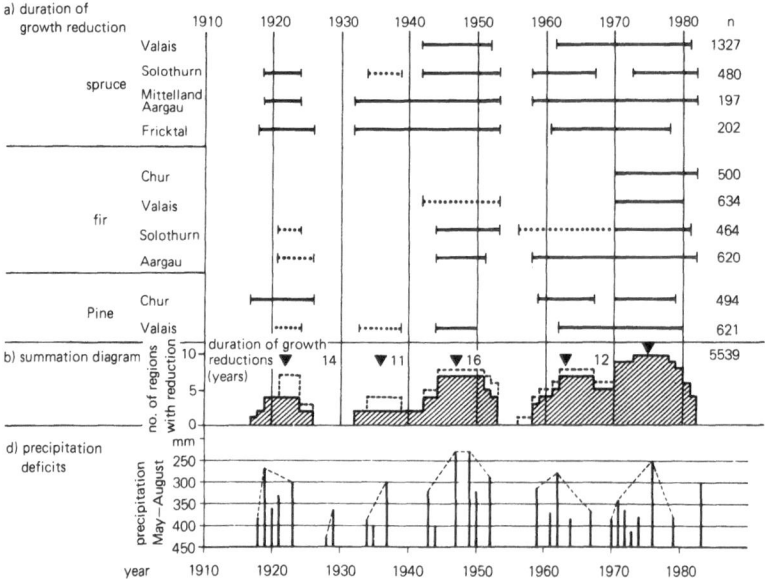

Fig. 9

Duration of growth reduction phases in different conifers (5539 trees) growing in different regions of Switzerland in relation to precipitation deficits in the months May-August as measured at the meteorological stations Rheinfelden, Olten and Aarau. The fluctuations are closely related to periods with low precipitation. After Schweingruber et al. (1986).

45

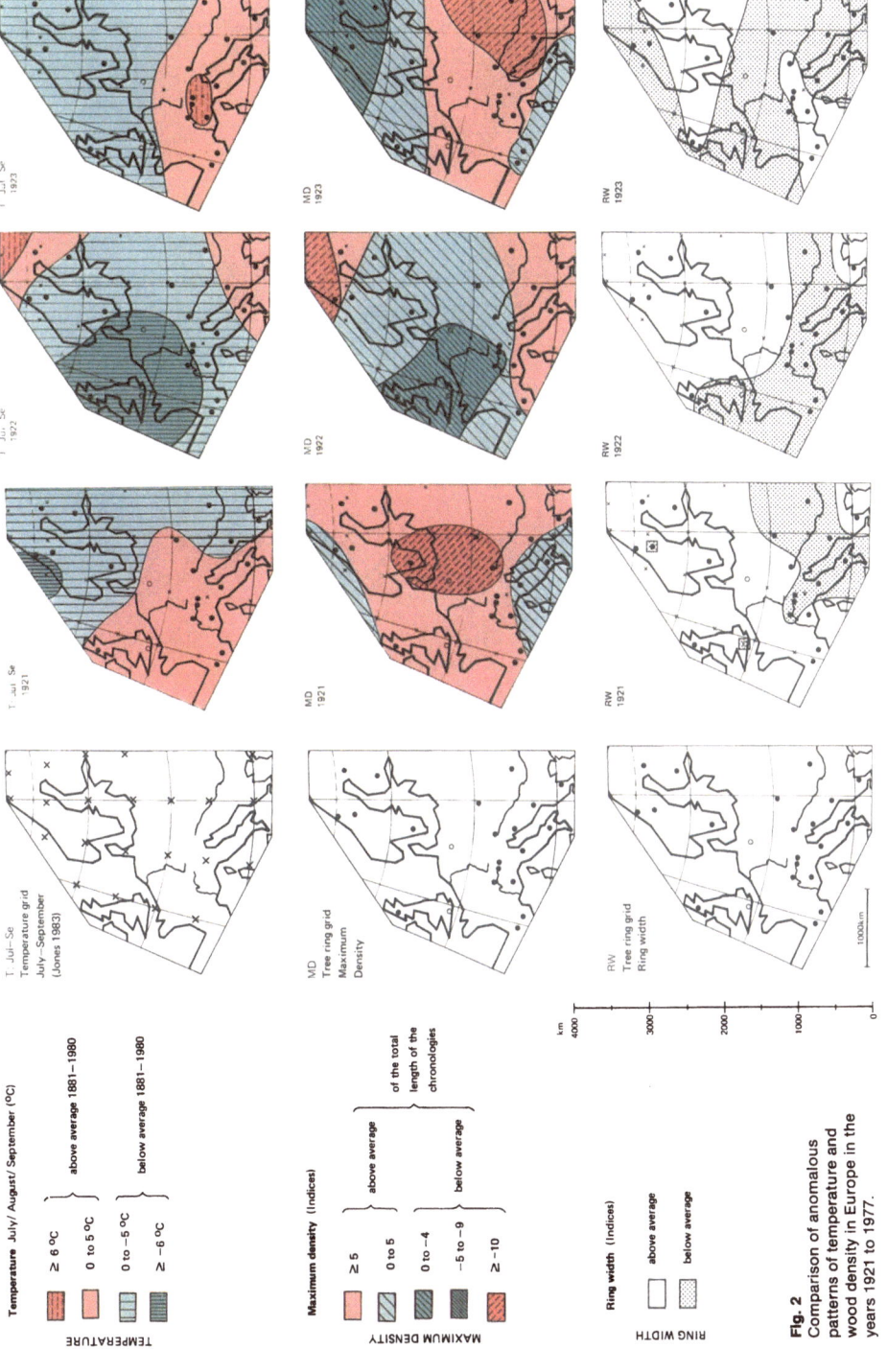

Fig. 2
Comparison of anomalous
patterns of temperature and
wood density in Europe in the
years 1921 to 1977.

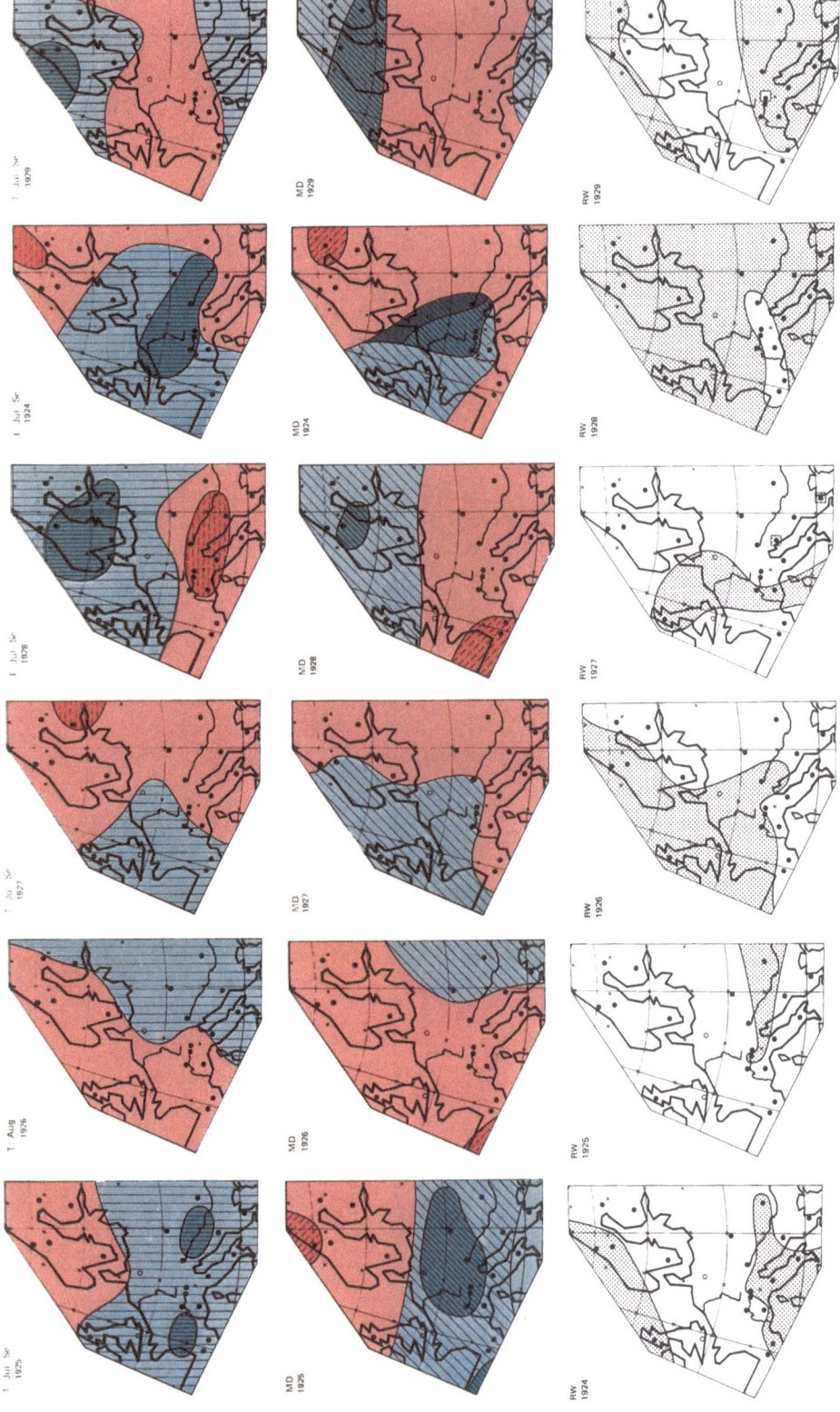

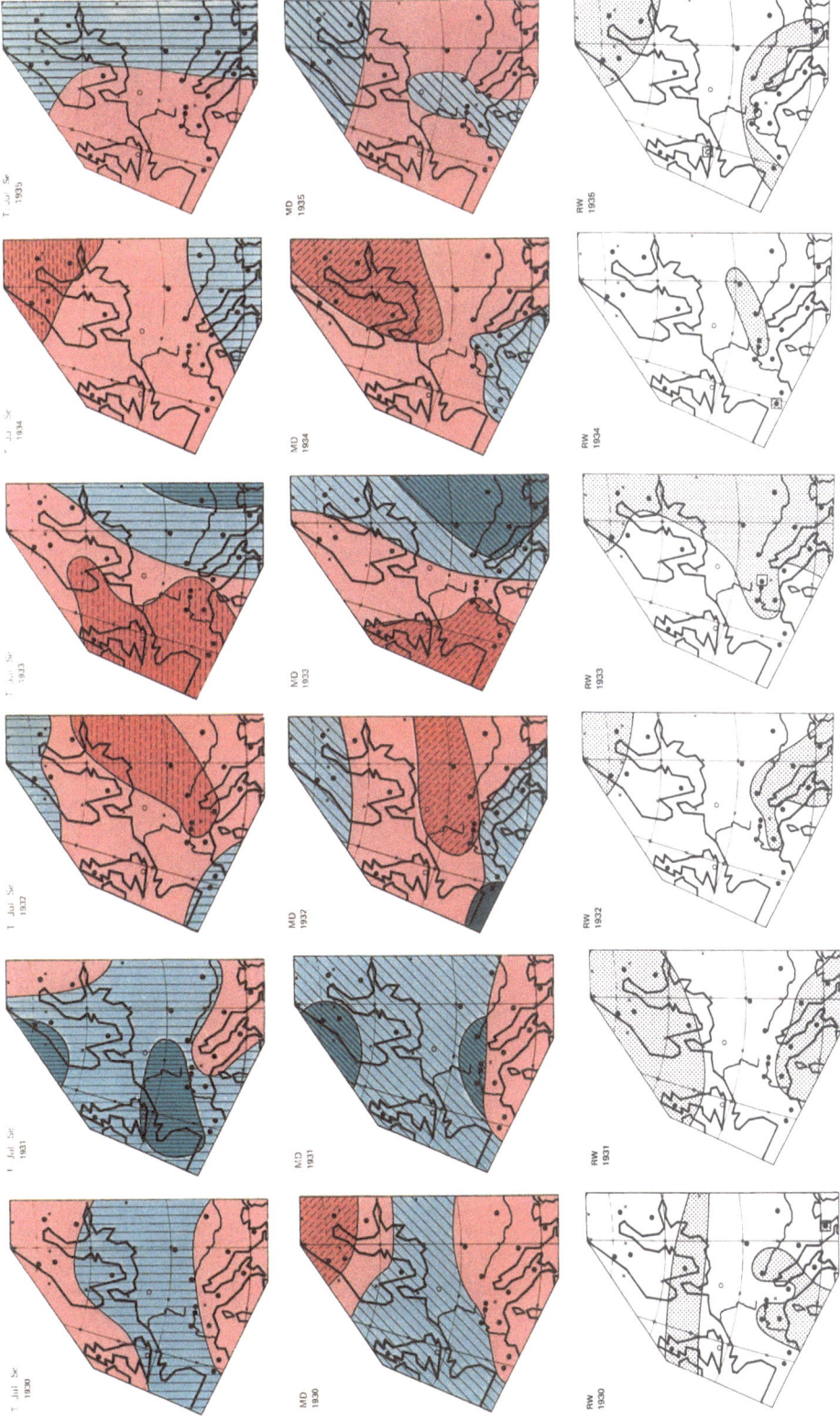

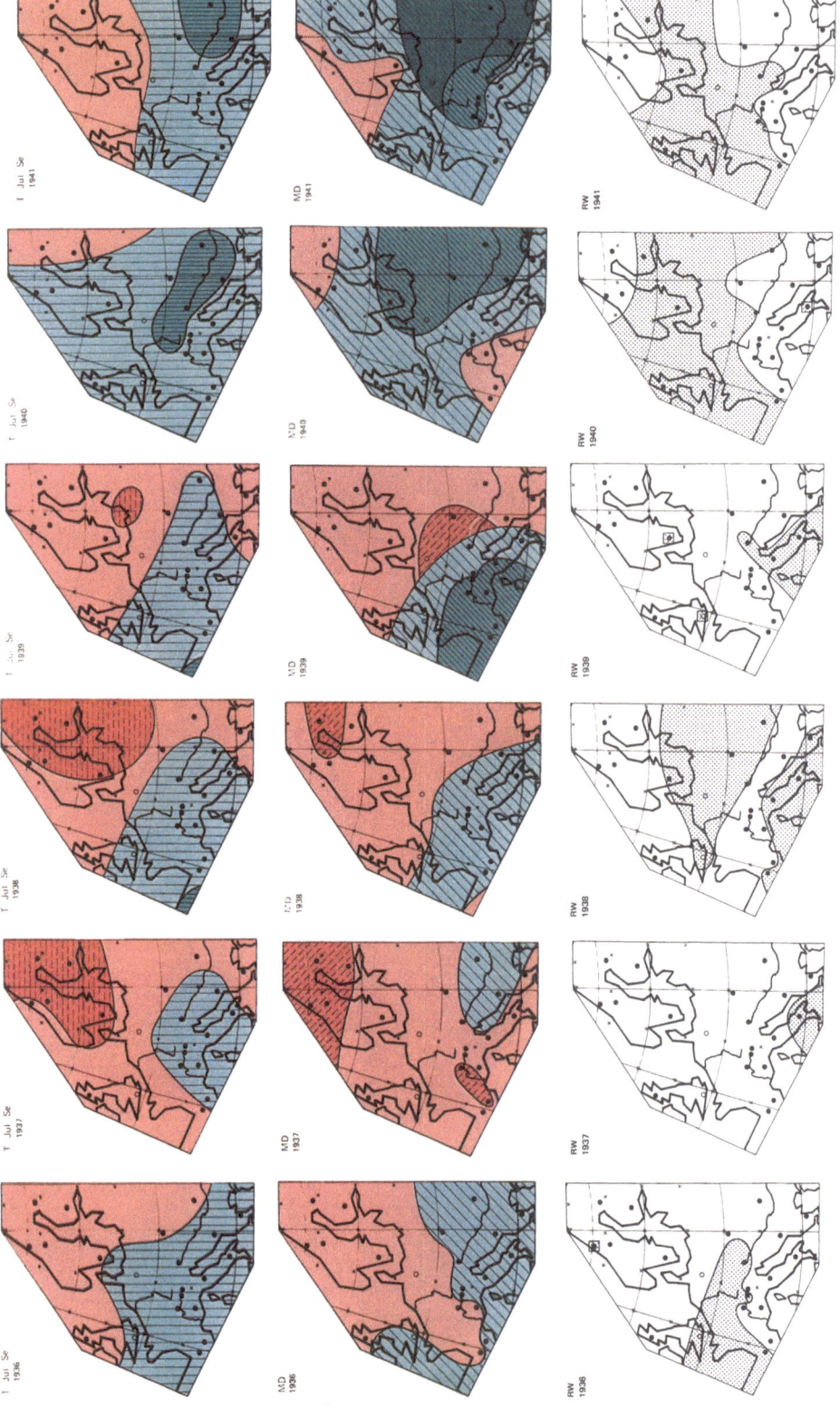

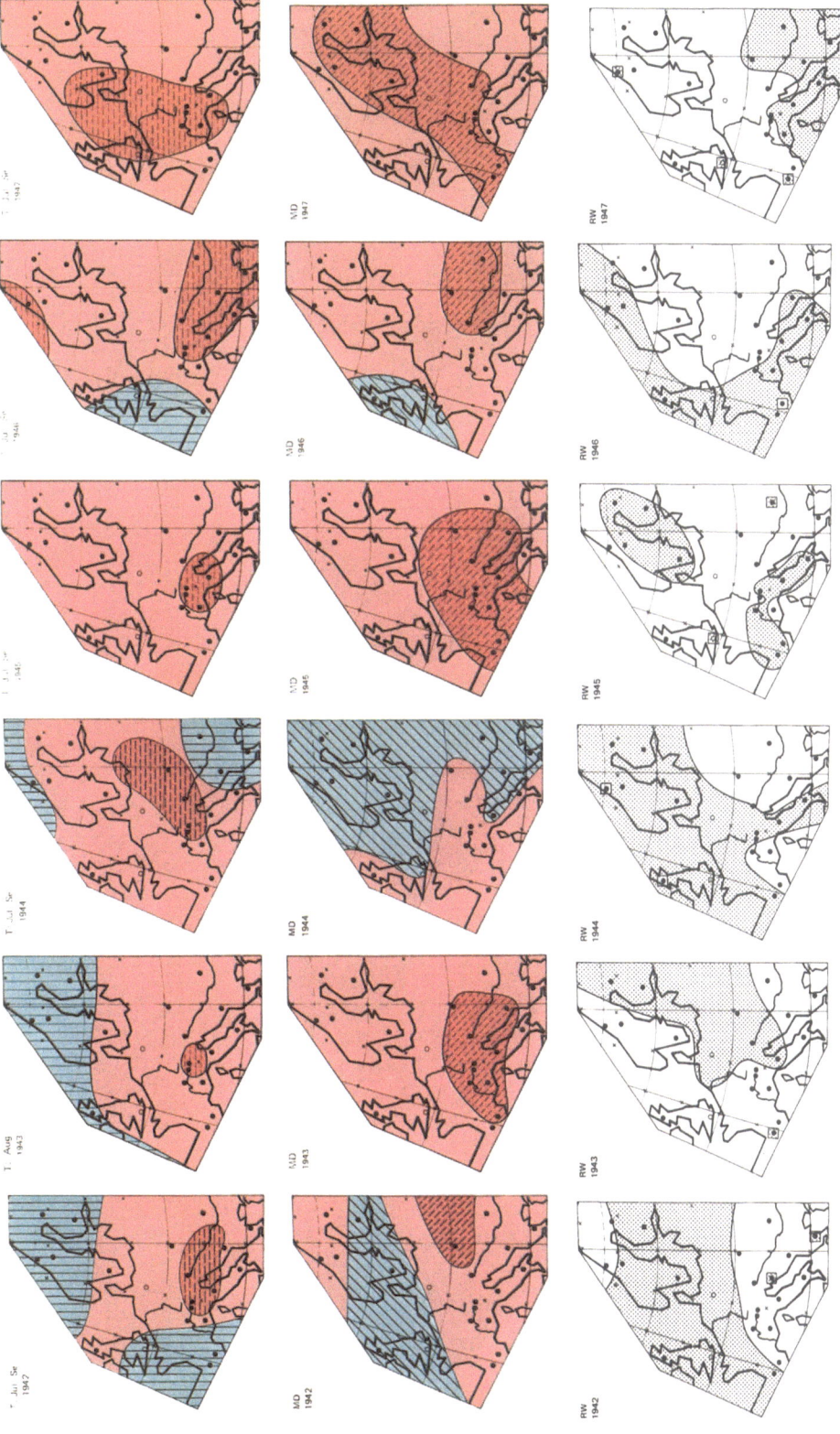

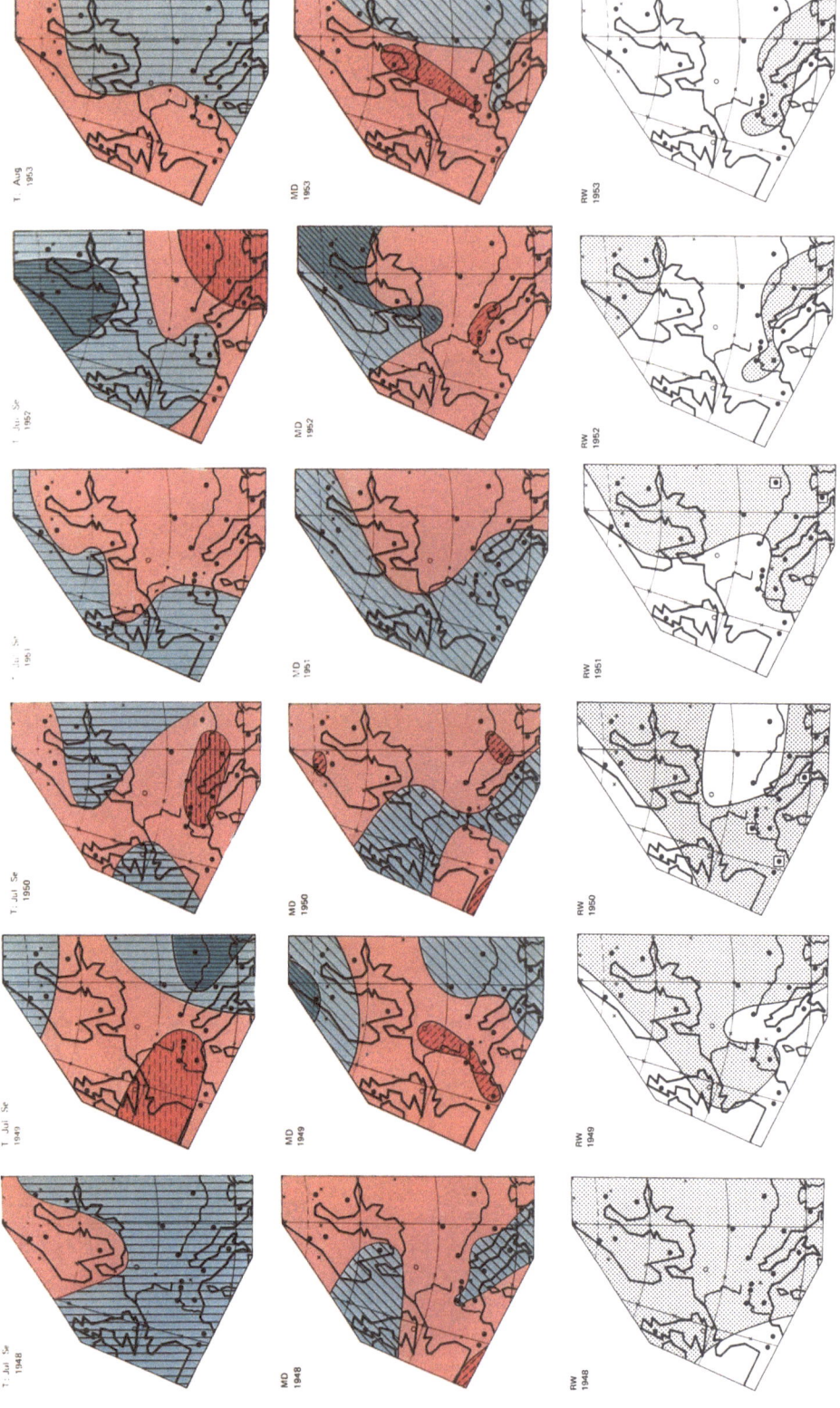

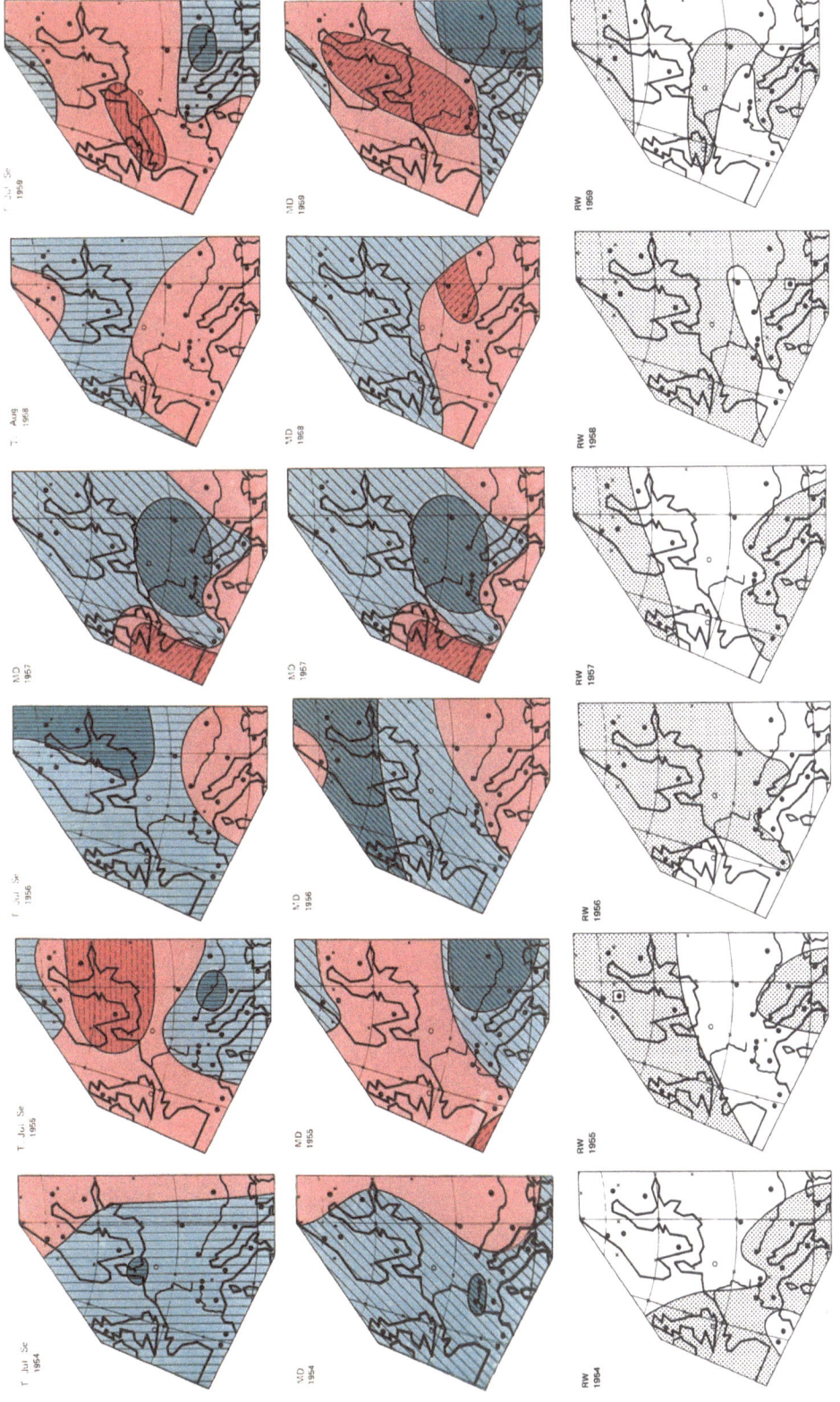

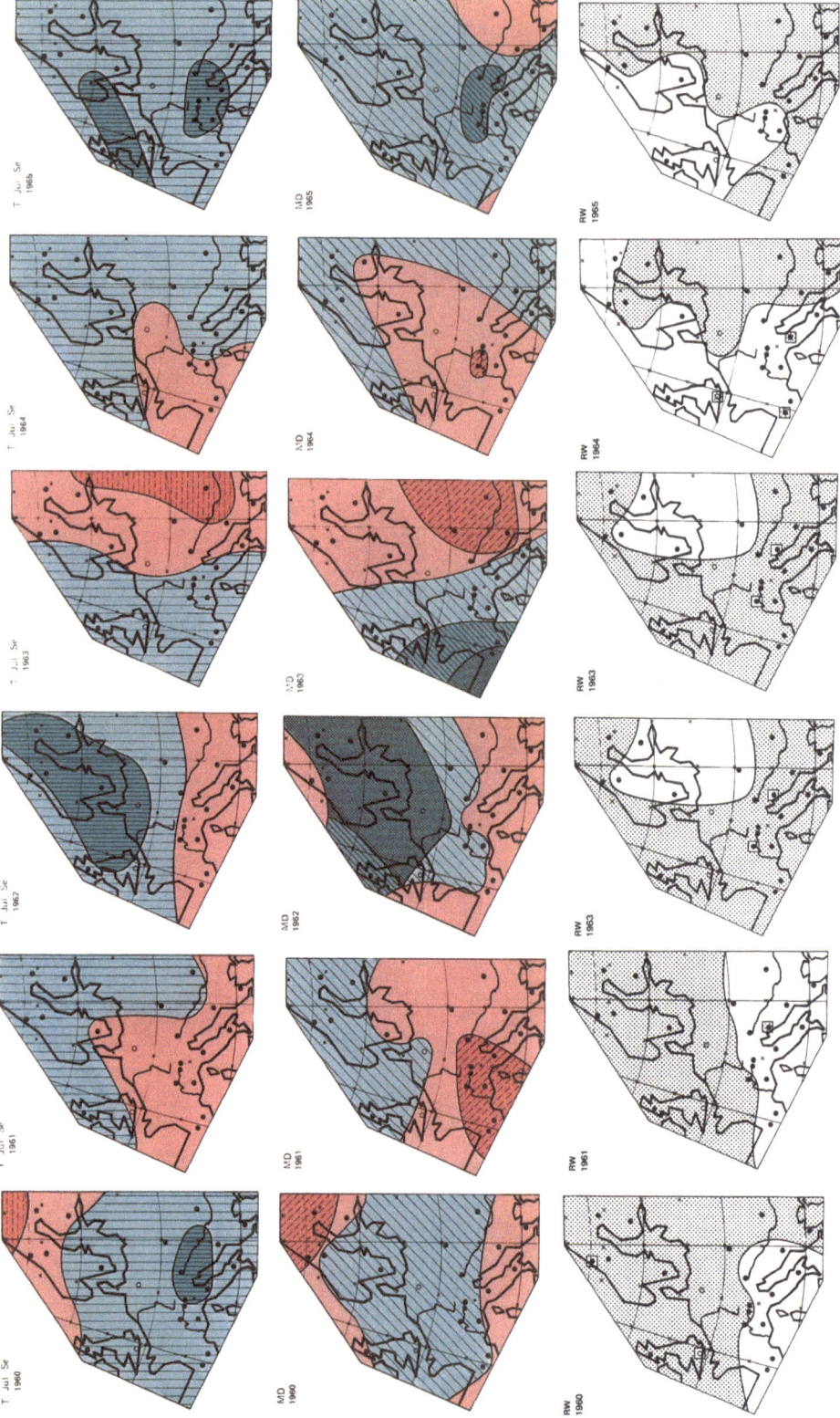

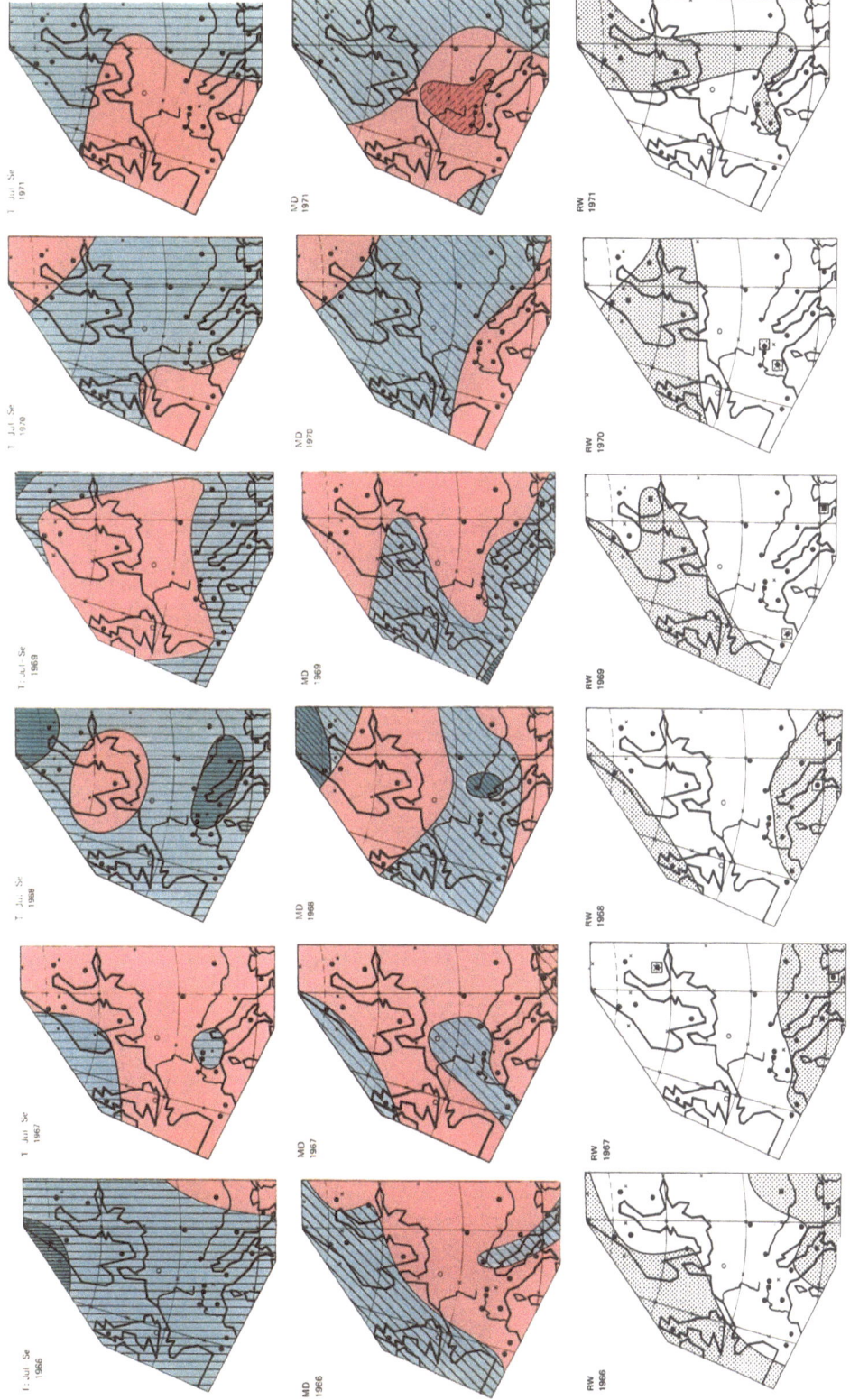

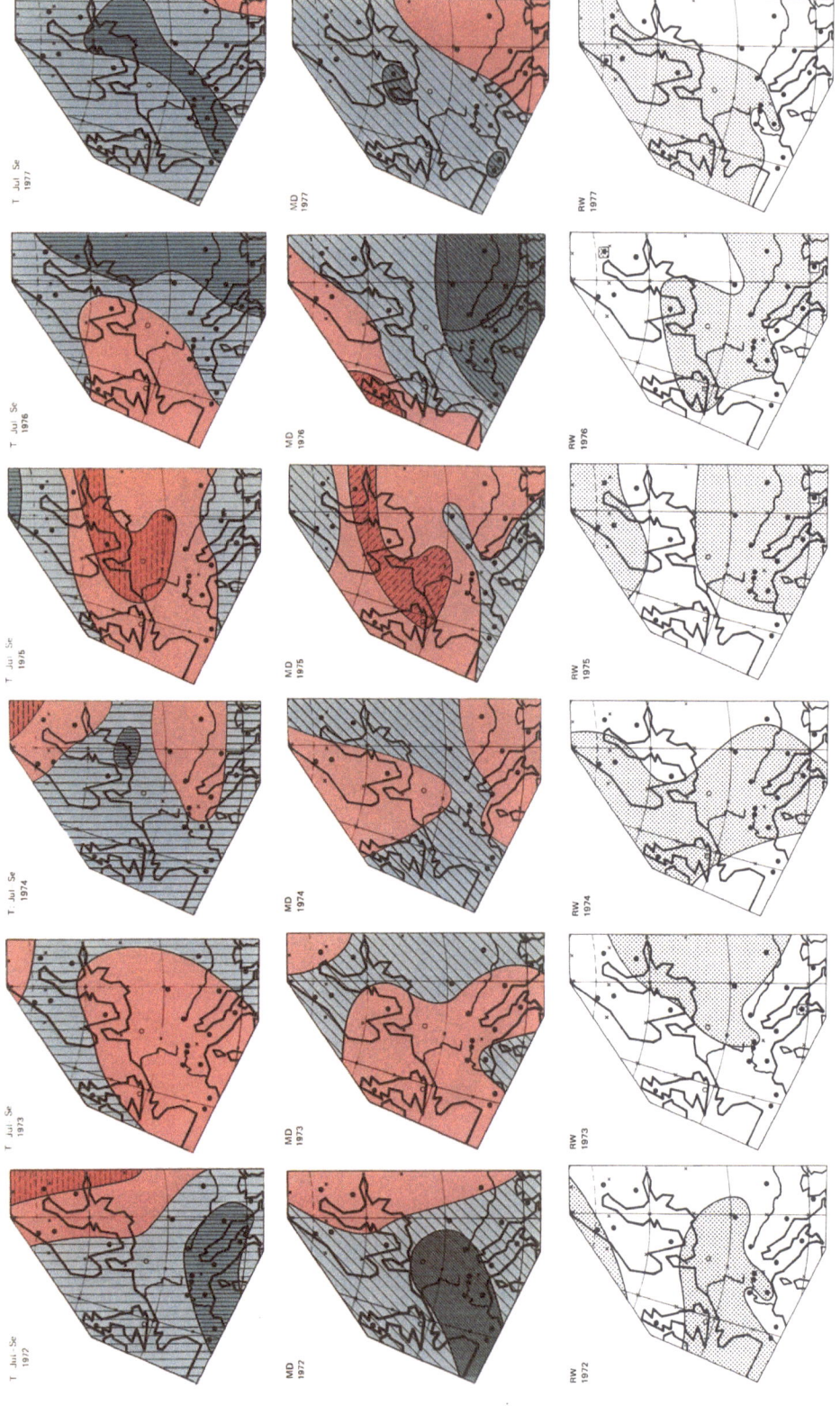

References

Fritts, H.C., 1974: Relationships of ring widths in arid-site conifers to variations in monthly temperature and precipitation. Ecological Monographies 44, 411-440.

Fritts, H.C., 1976: Tree rings and climate. London, New York, San Francisco. Academic Press, 567 pp.

Fritts, H.C., 1983: Tree-ring dating and reconstructed variations in Central Plains climate. Transactions of the Nebrasca Academy of Sciences, XL: 37-41.

Hughes, M.K., Schweingruber F.H., Cartwright, D., Kelly, P.M., 1984: July-August temperature at Edinburgh between 1721 and 1975 from tree-ring density and width data. Nature 308, 341-344.

Jones, P.D., Wigley, T.M.L., Kelly, P.M., 1982: Variations in surface air temperature: Part 1. Northern hemisphere 1881-1980. Monthly Weather Review 110, 59-70.

Kienast, F., 1985: Dendroökologische Untersuchungen an Höhenprofilen aus verschiedenen Klimabereichen. Ph.D. thesis, Univ. Zürich, 129 pp.

Kontic, R., Niederer, M., Nippel, C.A., Winkler-Seifert, A. 1986: Jahrringanalysen an Nadelbäumen zur Darstellung und Interpretation von Waldschäden (Wallis, Schweiz). Eidgenössische Anstalt für das forstliche Versuchswesen, Berichte 283, 1-46.

Lingg, W., 1986: Dendroökologische Studie an Fichte (Picea abies) und Weisstanne (Abies alba) im subkontinentalen Klimagebiet (Wallis, Schweiz) Eidgenössische Anstalt für das forstliche Versuchswesen. Berichte. 287, 1-81.

Parker, M.L., Henoch, W.S.E., 1971: The use of Engelmann Spruce latewood density for dendrochronological purposes. Canadian J. of Forest Res. 1, 90-98.

Schweingruber, F.H., Fritts, H.C., Bräker, O.U., Drew, L.G., Schaer, E., 1978: The X-ray technique as applied to dendroclimatology. Tree-Ring Bull. 38, 61-91.

Schweingruber, F.H., Albrecht, H., Beck, M., Hessel, J., Joos, K., Keller, D., Kontic, R., Lange, K., Nippel, C., Spinnler, A., Steiner, B., Winkler, A., 1986: Abrupte Zuwachsschwankungen in Jahrringabfolgen als ökologische Indikatoren. Dendrochronologia, 4, 125-183.

Shiyatov, S.G., Mazepa, V.S., 1986: Natural fluctuations of climate in the eastern regions of the USSR based on tree-ring series. Paper presented at the workshop on Regional Resource management. September 1985, Albena, Bulgaria. Collaborative Paper Internat. Inst. for Applied System Analysis. A-2361 Laxenburg, Austria. Vol. 1: 47-73.

Stockton, Ch.W., Mitchell, L.M., Meko, D.M., 1981: Tree-ring evidence of a relationship between drought occurrence in the western United States and the Hale Sunspot Cycle. In: LAWSON, M.P., BAKER, M.E., (eds.). The Great Plains. Perspectives and Prospects. University of Nebraska Press, Lincoln and London, 83-110.

VARIATIONS IN THE SPRING-SUMMER CLIMATE OF CENTRAL EUROPE FROM THE HIGH MIDDLE AGES TO 1850

Christian Pfister
University of Berne
Department of History
Engehaldenstrasse 4
3012 Berne/Switzerland

1. Does the climate of the High Middle Ages include elements for a warming scenario?

Warm periods in the past may provide elements for assessing the climatic and human consequences of the global warming which is predicted for the next century, if the present trend in concentration of greenhouse gases in the atmosphere continues (WMO, 1986). It is assumed that the sea ice around Greenland would retreat towards its northern coast in the early stage of a warming period and then completely disappear in a later stage. The Arctic Ocean would become ice free while the continental ice-dome at the Antarctic would persist. Such a situation existed during the Late Tertiary for the last time. Flohn (1984: 7, 265) concludes from the climatic evidence of this period that the northern coasts of the Mediterranean together with the Alps and south-central Europe (up to latitudes 48 - 50 N) might obtain a warm-temperate climate with some reduction of summer rains, i.e. with frequent warm season droughts, while the vegetation period would be increased by 1 - 2 months (Flohn, 1984: 9). On the other hand Frenzel concluded from the botanical evidence available from the warm interglacial periods over the past 700'000 years that the vegetation in Central Europe was not mediterranean at that time although summer temperatures may have been 2 - 3 degrees above the present average. This suggests a warm and moist summer climate.

What do we know about the warm period in the High Middle Ages? AD 985 Norse colonists from Iceland settled in Greenland around modern Narsaq, Julianehaab and Godthaab districts (Mc Govern, 1981: 407). The colonists were able to bury their dead deep in soil that has since been permanently frozen. In the mildest period in the early twelfth century the water in the fjords was at least sometimes 4° C, or more, warmer than the present normal (Lamb, 1982: 165 f). Drift ice reached the coasts of Iceland only on the average for a few weeks per year (Koch, 1945). In Central and Western Europe cultivation of the vine was spreading farther north, medieval vineyards in England are known up to a latitude of 53° N (Lamb, 1982: 170), reports on grape harvests from Bohemia, Thuringia and Belgium are included in medieval sources (Alexandre, 1986), vines were grown on altitudes of 600 to 700 m in the prealpine valley of Toggenburg (Scherer, 1874). In Norway, also, farm settlements were spreading up to 200 m higher than before on the hill country; wheat was grown almost to the latitude of the Polar Circle (Lamb, 1984: 36); in the Alps pastures could be grazed up to 2800 m (Röthlisberger, 1976). According to Lamb (1984: 37) midsummers during this "Little Optimum" were probably between 0.7 and 1.0° C warmer than the twentieth-century average in England and 1.0 - 1.4° C warmer in Central Europe (Lamb, 1982: 170).

A detailed analysis of the climate in the Middle Ages might therefore allow us to learn more on the seasonal weather patterns and on the anomalies that might be connected with the warming trend. This knowledge may be helpful for assessing the economic and societal impacts of a warming in the future. In the following the weather patterns in the spring-summer period between 1270 and 1425 will be investigated, and this data will be compared with the known variations in climate until the end of the so-called "Little Ice Age".

2.   Man-made data and their limitations

For the 350 years before the creation of the national weather service in Switzerland the monthly patterns of weather and climate could be described and quantified based upon a body of data that are mostly man-made. It comprises explicit weather data (early instrumental measurements, quantitative and qualitative descriptions of daily, weekly and monthly weather patterns) and proxy-data, i.e. a variety of information which reflects the combined effect of several weather factors during a period of several months (e.g. observations on the freezing of

lakes and the ripening of grapes and measurements of maximum tree-ring
density on logs from the upper timberline). The synchronous display of
all types of evidence in the CLIMHIST weather data bank (Pfister,
1985 a) has allowed to compare and to mutually check the different
types of data, to refine the interpretation and to derive monthly
indices for temperature and precipitation (Pfister, 1984).

Prior to the early sixteenth century man-made sources become at the
same time less abundant and less rich in meteorological entries. This
has two consequences: the time resolution of the reconstruction decreases, and the spatial dimension of the analysis must be increased.
The data are scattered within a large area, which begs the problem of
interpolation in space and reduces the reliability of the estimates, in
particular for precipitation. Also, continuous quantitative and homogeneous proxy-data that are required for estimating the temperature
patterns of the vegetative period are more difficult to obtain. Occasionally phenological observations have also been made in the Middle
Ages in order to determine and compare temperature patterns in outstanding years: a friar of the order of St. Dominic, who was born in
1221 and lived in Basel and in Colmar, has included phenological observations in his Annales Basilienses et Colmarienses. In 1283, one of the
earliest springs of the present millenium, he wrote for instance: the
first rye ears appeared around January 8th, the rye was in bloom on
March 19th, the vine got leaves on April 1st, the first new rye was
sold on May 17th, the peas could be harvested from June 8th, the same
date strawberries and cherries were ripe (Annales, 1861). But these
observations were not systematically carried on for some years, such as
those made in the eighteenth and nineteenth century. Thus they only
allow quantifying roughly the thermic character of climatic anomalies.

Grape harvest dates are available from the mid fourteenth century when
several chroniclers and annalists began to keep track of the date when
the wine harvest was fixed by public proclamation (Le Roy Ladurie,
1971: 50). However the records are often incomplete for many years.
Measurements of the maximum density of tree-rings at the upper timberline are the only continuous evidence for this time. The series of
Lauenen (Bernese Oberland) originates in 1269 (Schweingruber, 1978). In
a near future it will be extended back to the year 1000 (communication
by Dr. Schweingruber).

3.   Guidelines for the spatial extrapolation of data

Given the insufficient density of man-made and natural data in the
Middle Ages, it is essential to assess which biases might occur from
extrapolations within large areas. For this reason the spatial patterns
of temperature variation in Europe must be known for the present century. Moreover we need to know to what extent those patterns are bound
to change over time with the changing climate. For this reason the
analysis of spatial correlations should be extended back to the beginning of instrumental measurements. This type of analysis will be
attempted for Europe at the Geographical Institute of Berne based upon
a large number of long series readily available in machine readable
form (US Dept of Energy, 1985).

In the present context the spatial correlations of temperatures in the
vegetative period (April to September) and in summer are provided for
1901-60 and 1851-1900 (table 1).

If Zurich is chosen as a reference station the covariance of temperature patterns in the vegetative period is very high (R^2 of 65%) up to
the shores of the Atlantic over a distance of almost 800 kilometers and
still remarkable across the Alps to Northern Italy and to the Eastern
Alps (Vienna). The covariation between the summer temperatures (June-
August) is somewhat weaker in most cases. These results are in agreement with the significant correlations that have been found between
series of vine harvest dates over distances of 800 kilometers (e.g.
between Geneva and Vienna) (Flohn, 1985: 96).

4.   The use and misuse of historical sources

The meteorological evidence contained in the chronicles and annals of
the Middle Ages has been included in large compilations. At first sight
these compendia seem to provide a convenient ready-made data bank and
it is therefore not surprising that they have been much used by scientists seeking to reconstruct past climates. Historians in their turn
have drawn on the results of these reconstructions.

Table 1   Temperature correlations for summer months 1851-1900 and 1901-1960

First coefficient 1851-1900   Second coefficient 1901-1960

Kilometers: distance between the stations

a) April-Sept.:

	Nantes	Paris	Milano	Zürich	Munich
Paris	332km .88/.88				
Milano	832km .30/.43	624km .50/.68			
Zürich	752km .81/.78	480km .84/.90	210km .84/.70		
Munich	1005km .63/.65	690km .76/.82	345km .46/.68	255km .82/.94	
Vienna	1350km .41/.47	1020km .55/.67	630km .44/.55	585km .69/.81	360km .85/.88

b) June-August:

	Nantes	Paris	Milano	Zürich	Munich
Paris	.85/.87				
Milano	.18/.39	.30/.63			
Zürich	.66/.72	.74/.87	.79/.70		
Munich	.60/.60	.75/.80	.45/.66	.85/.94	
Vienna	.50/.43	.63/.66	.41/.50	.80/.78	.88/.86

Only a decade ago it was discovered that documentary sources of information about past climate are not equally reliable, and indeed much material which purports to record historical events is gravely misleading. As far as the Middle Ages are concerned, the current compilations have been analysed in detail by Bell and Ogilvie (1978). Their main weaknesses are inaccurate or uncertain dating of particular events, acceptance of accounts which are distortions or amplifications of original observations, inclusion of events for which there is no reliable evidence whatever and spurious multiplication of events through misdating. The consequences can be far-reaching. If for example a cold winter is misdated, which can easily occur, given that this season falls into two calendar years, this event may be included in a later compilation in an artificially multiplied way. Most fundamentally, the majority of works do not distinguish adequatly (if at all) between reliable and unreliable sources, and therefore they contain a mishmash of valuable and worthless data (Bell, Ogilvie, 1978). To take the well known Swiss compilation by Amberg (1890, 1892, 1897) as an example: for the medieval period half of the records are worthless. On the other hand Amberg did not include 30% of the reliable evidence that was available in print (Alexandre, 1986). Though the weaknesses of these compilations have been repeatedly demonstrated in the last years (Ingram et al., 1981: 192; Pfister, 1984: 40 f.) they are still uncritically used as data sources for climatic reconstruction (Burga, 1985), which is inacceptable. It is not necessary to comment on the value of sophisticated statistics that are based upon data from non contemporary sources (Pavese, Gregori, 1985).

Historians have been more cautious in selecting their sources. Schmitz (1968) has drawn from chronicles in order to investigate the links between meteorological variables and the prices for grain and wine from 800 to 1350. Buszello (1982) has illustrated the fluctuations in the standard of living of the "common man" in late medieval Switzerland, Baden and Alsace and their meteorological causes from 35 contemporary chronicles. A model of critical awareness is the recent work of the Belgian Pierre Alexandre (1986) who has brought together a new critical compilation of climatic evidence for Western and Central Europe up to 1425 (including Bohemia, Silesia and Northern Italy, but excluding England). He has taken care to assess the reliability of every source and to check every bit of information. All unreliable records were discarded. Alexandre only retained first hand observations from contemporary chroniclers. This is in itself an enormous task given the fact that most medieval sources only contain fragments of meteorological information. As far as man-made data are concerned this evidence will provide the basis for the following reconstruction.

5.   The representativity of tree-ring and grape harvest data

Tree-ring data from humid Western and Central Europe do not allow very convincing climatic reconstructions, mainly because of their long climatic memory (Hughes et al., 1982). Representative results can be expected from trees at the alpine timberline, where the temperature of the short vegetative period controls the growth rate. Significant progress has been made through the Roentgen density measurements of wood. Maximum density of the late wood is the single tree-ring charac- teristic most highly related to climatic data. A series from Lauenen (Bernese Oberland) originating in 1269 has been set up by Schweingruber (1978, 1979). Because in some years tree-ring density data are the only evidence available, their covariation with the thermal indices for summer (Pfister, 1984) had to be determined. For this purpose the original values were grouped into seven classes.It turned out that densities were very low in most of the very cold summers, whereas some of the hottest summers in the last 450 years (1616, 1719, 1947) do not stand out in the record (Pfister, 1985 c). This suggests that tree-ring density data should be used cautiously as climatic indicators unless they can be cross-checked with man-made observations or grape harvest dates.

One of the most representative series of grape harvest dates has gradually been built up by Le Roy Ladurie and Baulant (1980); they area-averaged 103 series of wine harvest data for eastern/central France, western Switzerland and a few villages from southwestern Germa- ny. The final series originating in 1484 was tested with the Parisian temperature series for the period 1797-1879. The coefficient of corre- lation is .86 which ought to reassure anyone to the reliability of phenological sources (Le Roy Ladurie, Baulant, 1980: 263). For the period before 1484 wine harvest data for Dijon are contained in the work of Angot (1883) from 1366 , who, in his turn, relied upon Lavalle (1855). The climatologist Jean-Pierre Legrand (1979 a, b) has used this evidence after 1400, when it is almost complete, in his careful investigation on temperature anomalies and sunspot activity over the last 580 years. New data for the fourteenth century have been discovered by historians who became sensitive to this type of evidence after reading Le Roy Ladurie's History of Climate (1967, 1971). Rotelli (1973) has included several series in his work on agrarian history of the Piemonte. The longest and the most complete (Moncalieri) covers the period from 1331 to 1424. A series from Beaune (Dubois, 1976) made it possible to bridge the frequent gaps contained in the Dijon series in the late fourteenth century. Another series from the plain of Albenga (N Italy) has been set up for 1364-1796 (Mazzei). Data for fifteenth century Anjou in form of a small graph (Le Mené, 1982) were discarded, because they were too difficult to check. For the period after 1371 a regional Côte d'Or series was computed from Dijon and Beaune; some missing values were interpolated using Albenga and Moncalieri (see Appendix). Previously the covariance between the four series was determined:

	Beaune	Dijon	Albenga
Dijon	.73 (N=30)**		
Albenga	.68 (N=22)*	.59 (N=18)*	
Moncalieri	.30 (N=32)*	.38 (N=42)*	.23 (N=26) ns

Significance: *≤.05   **<.00   N: paired observations
ns not significant

With r > 0.7 the correlation between the two series from the Côte d' Or is almost at the same level as in the later centuries (Le Roy Ladurie, Baulant, 1980: App. II). Remarkable also is the result of Albenga (across the Alps) while the covariance of Moncalieri with the Côte d'Or is weak and not even significant with Albenga.

The value of the Côte d'Or series for climatic reconstruction was critically assessed in cross correlating the series with the tree-ring maximum density series of Lauenen. In order to test the stability of the correlations, the series were split into two shorter periods. It turned out that the correlations between the two series were in the same order of magnitude as between the Lauenen series and the area-averaged series of wine harvest dates from Western Europe (Flohn, 1985: 98):

```
1370-1399:  -.65  (N= 29)
1400-1499:  -.43  (N= 94)
```

This result suggests that the Côte d'Or series can be used as a valid climatic indicator.

6. Outstanding anomalies

6.1. Definition and interpretation

In table 2 years with strong temperature anomalies in the vegetative period are listed from 1270 to 1524 as far as they appear in the tree-ring density record, in the series of wine harvest dates or in both. The two data sets complement each other: grape harvest dates primarily reflect temperatures in the spring-summer period (Pfister, 1984: 86; Flohn, 1985: 95 f.), maximum densities those in August and September (sometimes July through September), but also in April-May. However heat-waves in June, that may precipitately advance the maturation of grapes (Legrand, 1979 c: 43), as occurred e.g. in 1616 and in 1976, appear to have little effect on maximum density (Schweingruber, 1978: 78). Also the weather patterns in the Alps may be somewhat different than those in the lowlands. A more complete list of temperature anomalies may therefore be derived from a comparison of the two records and a cross-checking with additional unsystematic phenological observations and weather descriptions. Years in which the grape harvest began prior to September 10th or later than October 20th are considered anomalous, whereas for the tree-ring densities the limits of warm and cold anomalies are set to 1090 g/cm^3 and to 880 g/cm^3 respectively. A value above 1090 g/cm^3 corresponds in most cases to extremely high temperatures in August and September whereas densities below 880 g/cm^3 point to late summers that were colder than the chilliest of the present century (Pfister, 1985 c: 187).

Table 2   Temperature anomalies in the warm season 1270-1524 in
            Central and Western Europe

year	tree-ring density	wine-harvest dates	unsystematic phenological observations
1270	1170(+3)	-	first ripe grapes of early burgundy July 13 (Alsace)
1273	1180(+3)	-	-
1274	989(0)	Nov 18 (Basel)	
1282	1028(0)	new wine Aug 22 (Strasbourg)	-
1287	1124(+3)	end around Sept 22(Ribeauville)	-
1300	1138(+3)	-	-
1302	873(-3)	Oct 23(Limoges)	
1304	1090(+2)	-	first ripe grapes of early burgundy July 1st (Alsace)
1315	829(-3)	Nov 9 (Quimperlé) Nov 19 (Vienna)	-
1319	1123(+3)	-	-
1330	1075(+1)	Nov 9(Maillezais)	-
1331	1130(+3)	beg of Sept(Liege) Aug (Paris)	cherries ripe at beg of May (Maillezais)
1333	1169(+3)	-	-
1335	723(-3)	-	slow maturation of vine (Paris)
1336	1138(+3)	Aug(Liege)	very high sugar content (Zürich)
1345	831(-3)		slow maturation of vine (Paris, Torino)
1346	858(-3)		vine still in bloom on Aug 2nd (Lindau)
1347	909(-2)	Nov 9(Krems)	vine still in bloom on Sept 1st (Lindau)
1350	822(-3)	-	-
1359	724(-3)	-	-
1361	1010(+2)	Sept 9(Constance)	-
1366	778(-3)	Oct 17(Dijon)	slow maturation of vine (Mainz)

1370	870(-3)	-	slow maturation of vine (Mainz)
1382	1111(+2)	Sep 12(Dijon)	-
1383	1095(+2)	Aug (Rouen)Sept 5 (Bordeaux)	-
1384	1019(0)	Sep 9 (Beaune)	vine bloom begins on May 1st (St.Galler Rheintal)
1385	1063(+1)	Sep 8 (Beaune) Sep 9 (Dijon)	-
1393	1086(+2)	Sep 2 (Beaune) Sep 16(Dijon)	-
1400	1062(+1)	Sep 9 (Dijon)	-
1420	1003(0)	Aug 23(Paris) Aug 25(Dijon) Aug 29(Beaune)	new rye on May 15 (Mainz)
1422	1006(0)	Aug 28(Dijon) Sep 11(Beaune)	
1436	1028(0)	Oct 27(Dijon)	
1448	-	Oct 21(Dijon)	
1456	841(-3)	Oct 4 (Dijon)	
1465	787(-3)	Oct 12(Dijon)	
1473	1129(+3)	Aug 30(Dijon)	early rye harvest(Winterthur)
1480	890(-2)	Oct 10(Dijon) Oct 19(Lausanne)	
1481	952(-1)	Oct 18(Dijon)	
1488	1052(+1)	Oct 18(W Europe)	
1491	884(-2)	Oct 21(W Europe)	
1505	983(0)	Oct 14(W Europe)	
1511	958(-1)	Oct 15(W Europe)	

Sources: tree-ring data Lauenen: the series has kindly been
provided by Dr. Schweingruber
wine harvest dates and phenological observations up to
1426 (Alexandre, 1986)
wine harvest series of Dijon (Angot, 1883)
wine harvest series 'W Europe' (Le Roy Ladurie,
Baulant, 1980)

68

Graph 1: <u>The early spring-summer 1420 in Central Europe</u>

Data: Alexandre, 1986

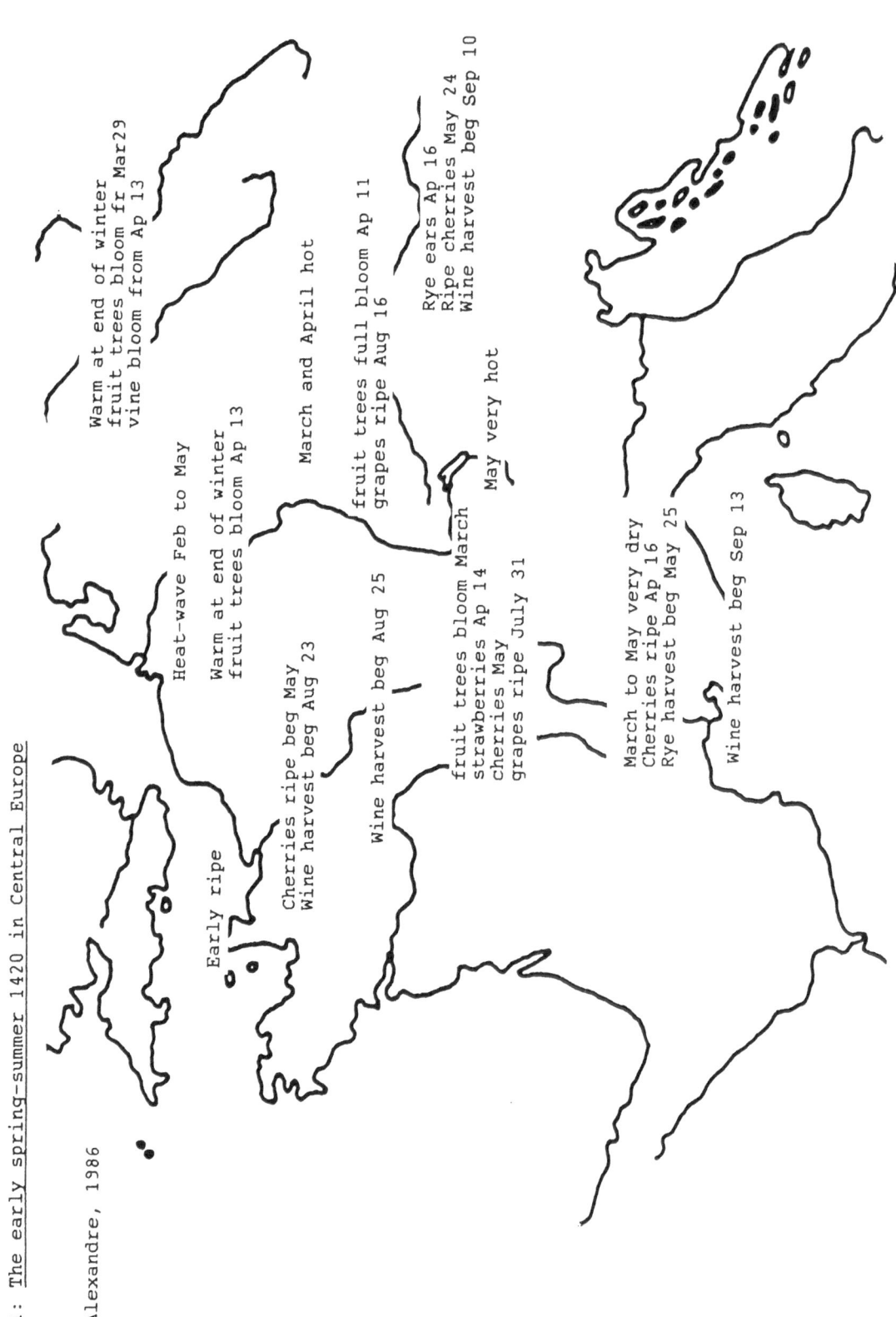

6.2. Examples of warm anomalies

In 1420 wine harvest in Western and Central Europe began at the end of August, even on altitudes of 500 to 700 m (Bern, Toggenburg). This is the earliest date ever recorded. Because the contemporaries considered this year outstanding, it was described in most chronicles, even in those in which meteorological observations were marginal. The anomaly extended from southern Thuringia to the Po valley and from Central France to the Vienna basin (graph 1). For the adjacent regions no data are available at present. In order to explain the weather patterns of this year the phenological evidence is compared to the pattern observed in 1540 (Pfister, 1984, 1985 a) and to comparable phenological extremes documented with thermometrical measurements (table 3).

In 1420 the warm phase started in February. In March summer began already. The vine bloom was two weeks earlier than in 1893 - the most advanced year within the instrumental period. Based on the date for Lichtensteig (600-700 m) and according to a gradient of 3.6 days per 100 m (Becker, 1969 : 142) it has been estimated that the end of the bloom may have occurred in the last days of May around Basel (260 m) i.e. almost a month before the mean date of the present century. The first new wine ("Sauser"), probably from early burgundy grapes, was sold at the beginning of August. The wine harvest was advanced by a month compared with the long term average for Western Europe (Le Roy Ladurie, Baulant, 1980). In 1540 the heat-wave began in April (as in 1893). It is reported that the development of the vegetation was slowed down by drought. The meager evidence available for 1270, 1304, 1331 and 1336 (cp. table 2) suggests that phenological patterns may have been comparable to those observed in 1420 and in 1540: in 1270 and 1304 the early burgundy grapes were ripe at about the same time as in 1540, whereas in 1331 the ripening of the first cherries in Western France and the beginning of the wine harvest in Paris coincided roughly with the corresponding phenophases in 1420.

The comparison of the phenophases in 1420 and 1540 with the corresponding extremes documented with thermometrical evidence suggests that in 1420 all months from February to August (in 1540 from April to August) may have been 2 to 3 degrees above the 1901-60 average.

Table 3   Pattern of the phenophases in Southern Central Europe in 1420 and 1540
compared to the earliest known phases within the instrumental period

	1420	1540	within instrumental period	cumulative temperature deviations (°C)
Trees in bloom	March (Berne)	-	March 29,1948 (Hallau)	+ 9 (Jan. to March)
Ripe Strawberries	Apr. 23 (Basel)	-	-	
Vine in bloom	May 2 (Basel)	-	May 17,1893 (Hallau)	+ 4 (Apr., May)
Ripe cherries		beg.June (Zürich)		
Bloom ends	June 6 (Lichtensteig)	June 10 (Schaffhausen)	June 13,1811 (Schaffhausen)	+ 6 (Apr. to June)
New rye	May 15 (Metz)	June 16 (Zürich)	June 30,1822 (Schaffhausen)	+ 6.5 (May, June)
Grapes ripe		July 10 (Zürich)		
New wine	July 31 (Metz) Aug. 10 (S.Baden)	Aug. 15 (Zürich)		
Wine harvest beg.	Aug. 28 (Lichtensteig) Aug. 31 (Berne)	Sep. 9 (Zürich)	Sep. 9,1822	+ 6.5 (May, June)

Sources: Pfister 1984, 1985a; Alexandre 1986.

Precipitation patterns may only be got for 1540. Heinrich Bullinger, who was antistes in Zurich, recorded a total of six days with precipitation during the 26 weeks from mid March to the end of September. After two rainy days at the beginning of October the weather turned to warm and dry again. On New Year boys were still swimming in the Rhine near Schaffhausen. The record of this outstanding year suggests that a mediterranean type of climate persisted for about ten months. Annual precipitation may not have exceeded 300 to 400 mm (Pfister, 1984: 138). The summer of 1304 may have been similar. According to the Annales Colmarienses flour became scarce because many mills fell dry; wine was abundant, but the casks couldn't be loaded on boats because the level of the Rhine was too low (Annales, 1861: 231).

The most outstanding spring-summer period since 1269 probably is 1473, because in this year a very early grape harvest coincided with a tree-ring density that is close to the maximum recorded (cp table 2). There is an abundance of observations for this anomaly, but the documentation is not available yet.

6.3. The ice-age summers of the 1340's

From the climatic history of the last centuries it is well known that cold summers have a certain tendency to cluster (e.g. 1812 to 1817). A similar pattern stands out in the 1340's: for 1345 a slow maturation of the vine is reported from Paris and Torino, in 1346 the vine was still in bloom at Lindau after August 2nd, a retardation of vegetative growth that may be compared only to the two coldest summers since 1500 (1628 and 1816). In 1345 and 1346 maximum tree-ring densities are among the twenty lowest contained in the Lauenen series. In 1347 the vine was still in bloom at the beginning of September. This points to a cold anomaly in July and August that is unique in the last six centuries (Pfister, 1985 c: 192).

7. Climatic trends in spring and summer from the High Middle Ages to 1850

For the warm period of the High Middle Ages continuous proxy-data are not yet available. A historical custom introduced by the order of Cluny in 1018 nevertheless allows estimating the average date of the grape harvest in Northern and Central France. At the mass of the Transfiguration (Aug. 6) the new wine was dedicated at the altar and afterwards presented to every friar. When the maturation of the vine was delayed, the juice of some soft grapes was taken instead. If the correction for the Gregorian rule is made the average date of this celebration was around August 13. This roughly corresponds to the earliest date of the wine harvest ever recorded (August 13, 1893 in the region of Bordeaux). In this year the mean temperature from April to August was 2.6 degrees above the long term average (Legrand, 1979 b: 42 f.). But presumably spring-summer temperatures in the High Middle Ages were somewhat lower. Legrand admits that we do not know whether the grapes for the first new wine were grown on an espalier sheltered from the cold. Also it must be assumed that the wine was made from an early variety of burgundy grapes. In Switzerland the early burgundy grapes were ripe around August 10 in very warm summers (Pfister, 1984: 84). In 1420 the first "new wine" from these grapes was drunk at the beginning of August, in 1540 on August 15 (table 3), about twenty days before the wine harvest was opened. If this delay was the same in ordinary years, the mean grape harvest date would be around September 1st in the High Middle Ages, which is a few days earlier than in the warmest summers documented with thermometric measurement. From a regression approach comparing the decennial means of wine harvest dates and temperatures from 1370 to 1850 it has been estimated that an opening of the harvest on September 1st corresponds to a mean temperature from April to September that is 1.7 (+ - 0.2) degrees above the average for 1901-60.

From the spreading of the vine and the cereals to higher altitudes and latitudes in the High Middle Ages it has been primarily concluded to temperature patterns in midsummer. However, as the maturation of both crops is mainly promoted by temperatures in May and June (Pfister, 1984: 86) this evidence is rather conclusive for conditions in late spring and early summer (Legrand, 1979 b: 43). An advance in the mean date of the grape harvest also suggests an earlier date of snow-melt in the Alps since phenophases of the vine are significantly correlated with the melting dates at different levels of altitude (Pfister, 1985 b: 168 f.).

In the Lauenen series of tree-ring densities the warm period of the
High Middle Ages can be documented by assessing the frequency of
positive anomalies exceeding 1089 g/cm^3.

```
1269 - 1299: 13 %    1300 - 1339: 18 %
1340 - 1399:  5 %    1400 - 1499:  1 %
1500 - 1599:  2 %    1600 - 1699:  0 %
1700 - 1799:  5 %    1800 - 1979:  0 %
```

Within the entire series they account for 3% of the cases. In the four
years 1774, 1777, 1779, 1781, the only ones documented by thermometrical measurement, temperatures in August and September were 2.0 degrees
above the 1901-60 average. From 1269 to 1339 positive anomalies occurred more than once every decade on average; this suggests that they
were part of the "normal" climatic pattern; after 1340 their frequency
drops to a level of 5%, after 1400 they became very rare. Not a single
occurrence is measured for the seventeenth century. Since 1781 they
have not been recorded any more. Undoubtedly the early fourteenth
century marks a climatic watershed. The frequent occurrence of high
maximum densities before 1330 can be interpreted in the context of a
warm climate to which an advance in the beginning of the grape harvest
is connected. It may be hypothesized that summers which were outstanding according to the standards of later periods, such as those of
1420, 1473 or 1540 were within the normal range of fluctuations during
the High Middle Ages. This could explain why extreme anomalies in the
warm period such as the summer of 1331 are only briefly described in a
few chronicles whereas similar events after the mid-fourteenth century
have evoked extensive comments in a multitude of sources. The evidence
on precipitation does not contradict this hypothesis. From 1200 to 1310
only two decades had a moderate excess of wet summer months (graph 2).

Graph 2 <u>Precipitation patterns in Summer (J, J, A) from 1150 to 1420</u>
Difference of unmistakably rainy and dry months per decade

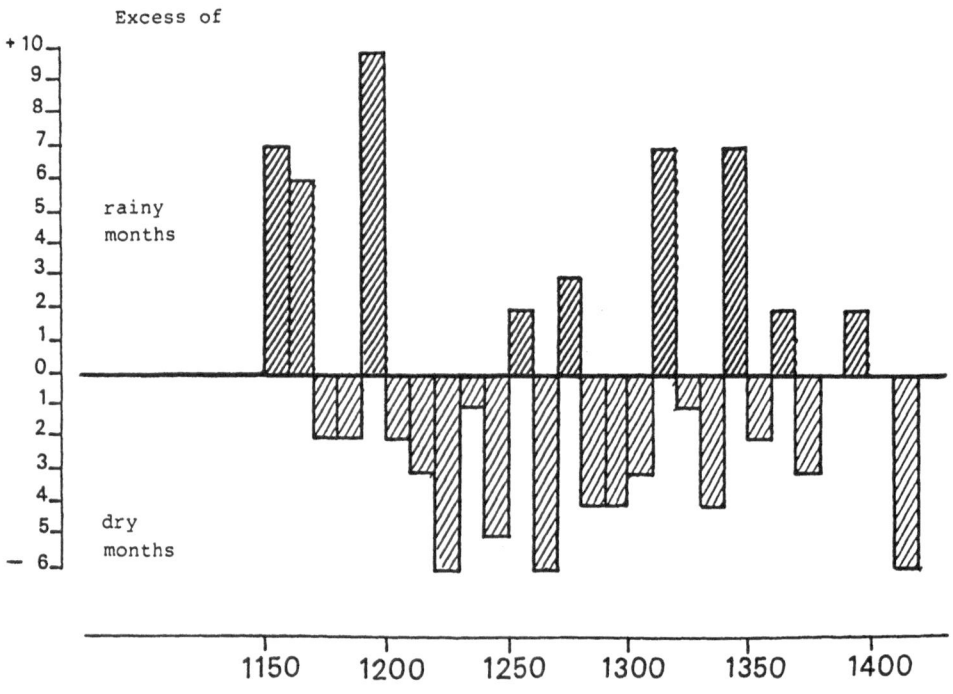

The year indicates the beginning of the decade

Source: Alexandre, 1986

The frequency of the very cold summers can be documented from the Lauenen series by assessing the frequency of negative anomalies below $880 \ g/cm^3$.

1269 - 1299:	0 %	1300 - 1339:	5 %
1340 - 1379:	18 %	1380 - 1429:	0 %
1430 - 1499:	3 %	1570 - 1599:	17 %
1600 - 1699:	4 %	1700 - 1810:	0 %
1811 - 1860:	10 %	1861 - 1979:	0 %

The evidence suggests that the shift from the warm climate of the High Middle Ages to the full brunt of the "Little Ice Age" did not take much more than two decades. The end of the transitory period in the 1330's stands out by an extreme variability: five late summers out of ten were either much too cold or much too warm: three of them (1331, 1333, 1336) are very close to the highest tree-ring density recorded, whereas the value of 723 g/cm^3 measured for 1335 is the lowest within the entire series. The frequency of very cold summers from 1340 to 1379 may be compared to the final decades of the sixteenth century and to the period 1811 to 1860, which were at the origin of far reaching glacier advances in the Alps. For the Aletsch glacier an advance of a similar magnitude in the late fourteenth century has been recently demonstrated by Holzhauser (1984).

In order to compare the spring-summer climate in the Late Middle Ages with that of more recent periods the series of wine harvest dates for the Côte d'Or has been joined to the series for Western and Central Europe (Le Roy Ladurie, Baulant, 1980). The curve of the decennial means of grape harvest dates (graph 3) is based upon the Côte d'Or series until 1500 and on the other series for the following period. The differences between both series in the overlapping period may be neglected.

Within the series two levels may be clearly distinguished: from 1380 to 1430 the beginning of the wine harvest was advanced six days compared to the long term average; from about 1450 to the late 19th century the curve fluctuates around a lower level and displays the known advances of glacier history: the late sixteenth century, the 1690's, the 1740's, the 1770's, the 1820's. The small trough in the 1490's may have preceded a minor advance of the Lower Grindelwald Glacier, which, in 1535, was very close to the valley (Pfister, in preparation). The advanced development of the vegetation from 1380 to 1430 may be explained by the complete absence of ice age summers and by an enhanced frequency of warm spells in the spring-summer period (e.g. 1382-1385). In the 1380's and the 1420's temperatures from April to September may have been 0.5 degrees above those of our century. A rapid and prolonged melting back of the alpine glaciers may be hypothesized for this period.

Graph 3   Mean date of grape harvest in Western Europe 1370-1880

Decennial averages. The years give the beginning of
the decade (e.g. 1500 = 1501-1510)

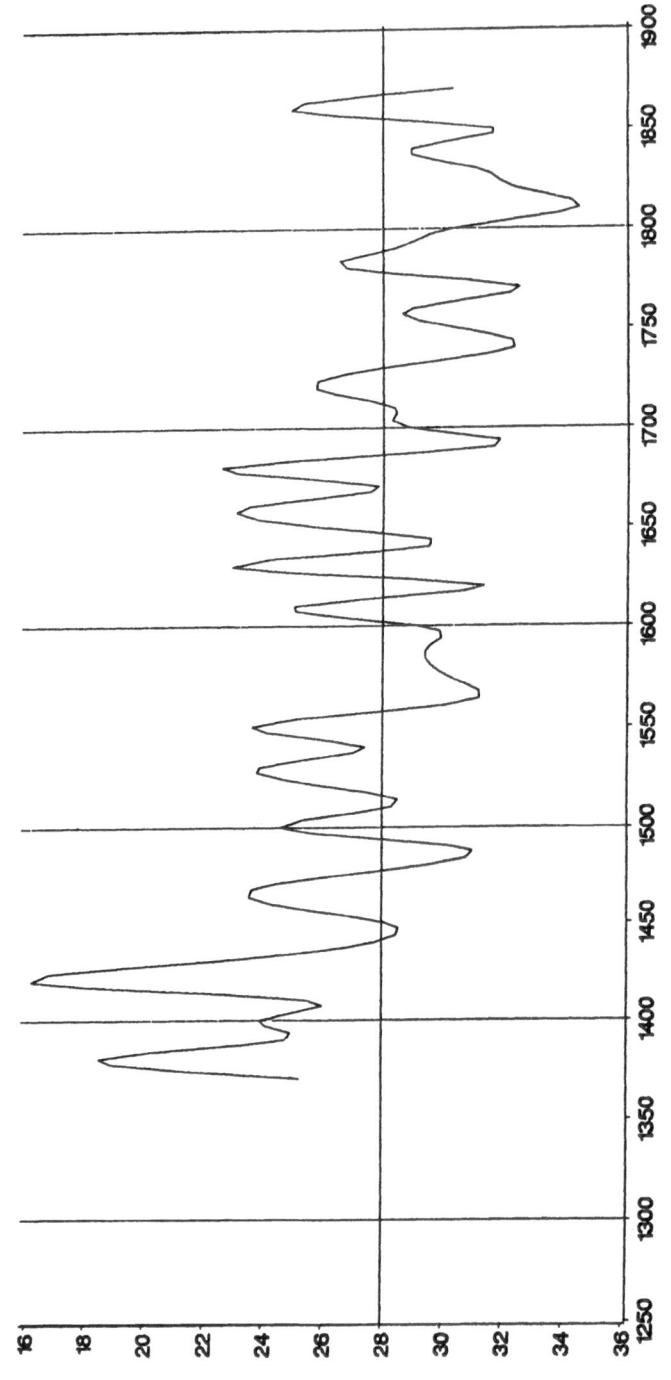

days from
Sep 1st(=1)

decade

Source: 1370 to 1500   Series Mont d'Or
        1500 to 1880   area-averaged series W Europe (Le Roy
                        Ladurie, Baulant, 1980)

APPENDIX   -   GRAPE HARVEST DATES   (DAYS FROM SEPT. 1ST = 1)
ORIGINAL AND INTERPOLATED SERIES

YEAR	ALBENGA	MONCALIERI	AVIGNON	BEAUNE	DIJON	COTE-D'OR	W.-EUROPE
1370	.	25	.	.	.	28M	.
1371	.	23	.	.	.	26M	.
1372	.	26	.	27	.	29	.
1373	.	18	.	20	.	22	.
1374	.	26	.	31	.	33	.
1375	.	32	.	20	.	22	.
1376	20	28	.	16	.	18	.
1377	22	20	.	.	.	22A	.
1378	.	26	.	23	.	25	.
1379	.	26	.	22	.	24	.
1380	.	23	.	21	.	23	.
1381	26	30	.	21	20	21	.
1382	.	21	.	13	12	13	.
1383	16	34	.	5	.	7	.
1384	5	19	.	6	.	8	.
1385	19	13	.	7	9	8	.
1386	24	29	.	18	.	20	.
1387	24	29	.	20	33	27	.
1388	15	29	.	28	25	27	.
1389	27	24	.	24	24	24	.
1390	20	29	.	.	.	20A	.
1391	25	29	.	17	.	19	.
1392	30	22	.	38	33	36	.
1393	17	14	.	1	16	9	.
1394	22	32	.	.	38	38	.
1395	21	.	.	.	.	21A	.
1396	.	25	.	.	.	27M	.
1397	23	23	.	.	22	22	.
1398	.	.	.	.	25	25	.
1399	29	.	.	.	26	26	.
1400	13	26	.	.	10	10	.
1401	23	16	.	14	11	13	.
1402	.	25	.	13	20	17	.
1403	25	29	.	21	19	20	.
1404	25	29	.	28	.	30	.
1405	23	.	.	26	.	28	.
1406	23	.	.	24	.	26	.
1407	20	22	.	25	29	27	.
1408	23	.	.	27	33	30	.
1409	24	23	.	17	30	24	.
1410	.	.	.	10	24	17	.
1411	.	.	.	38	37	38	.
1412	.	.	.	17	18	18	.
1413	20	18	.	10	29	20	.
1414	.	.	.	27	33	30	.
1415	.	22	.	20	22	21	.
1416	.	.	.	24	31	28	.
1417	.	.	.	20	22	21	.
1418	.	.	.	9	14	12	.
1419	.	.	.	17	25	21	.
1420	12	20	.	-3	-6	-5	.
1421	.	.	.	23	22	23	.
1422	.	.	.	10	-3	4	.
1423	.	.	.	22	23	23	.
1424	.	.	.	10	11	11	.
1425	.	.	.	.	16	16	.
1426	.	16	.	24	14	19	.
1427	.	.	.	.	25	25	.
1428	.	.	.	.	36	36	.
1429	.	.	.	.	24	24	.
1430	.	.	.	.	15	15	.
1431	.	.	.	.	19	19	.
1432	.	.	.	.	18	18	.
1433	.	.	.	.	12	12	.
1434	.	.	.	.	1	1	.
1435	.	.	.	.	25	25	.
1436	.	.	.	.	56	56	.
1437	.	.	.	.	28	28	.
1438	.	.	.	.	.	.	.
1439	.	.	.	.	27	27	.
1440	.	.	.	.	28	28	.
1441	.	.	.	.	18	18	.
1442	.	.	.	.	13	13	.
1443	.	.	.	.	26	26	.
1444	.	.	.	.	22	22	.
1445	.	.	.	.	36	36	.
1446	.	.	.	.	.	.	.
1447	.	.	.	.	.	.	.

INTERPOLATION : 'A' FROM ALBENGA
'M' FROM MONCALIERI

APPENDIX   -   GRAPE HARVEST DATES
ORIGINAL AND INTERPOLATED SERIES

YEAR	ALBENGA	MONCALIERI	AVIGNON	BEAUNE	DIJON	COTE-D'OR	W.-EUROPE
1448	.	.	.	.	50	50	.
1449	.	.	.	.	29	29	.
1450	.	.	.	.	25	25	.
1451	.	.	.	.	39	39	.
1452	.	.	.	.	24	24	.
1453	.	.	.	.	27	27	.
1454	.	.	.	.	34	34	.
1455	.	.	.	.	31	31	.
1456	.	.	.	.	33	33	.
1457	.	.	.	.	14	14	.
1458	.	.	7	.	18	18	.
1459	.	.	28	.	36	36	.
1460	.	.	18	.	.	22V	.
1461	.	21	13	.	16	16	.
1462	.	13	5	.	.	9V	.
1463	.	.	33	.	35	35	.
1464	.	.	20	.	14	14	.
1465	.	.	35	.	41	41	.
1466	.	.	28	.	27	27	.
1467	.	.	.	.	27	27	.
1468	.	.	.	.	32	32	.
1469	.	28	.	.	20	20	.
1470	.	.	.	.	37	37	.
1471	.	.	.	.	11	11	.
1472	.	.	.	.	23	23	.
1473	.	.	.	.	-2	-2	.
1474	.	.	.	.	39	39	.
1475	.	.	.	.	31	31	.
1476	.	.	.	.	28	28	.
1477	.	.	.	.	41	41	.
1478	.	.	.	.	19	19	.
1479	.	.	.	.	16	16	.
1480	.	.	.	.	39	39	.
1481	.	.	.	.	47	47	.
1482	.	.	.	.	16	16	.
1483	.	.	.	.	15	15	.
1484	.	.	.	.	20	20	31
1485	.	.	.	.	43	43	37
1486	.	.	.	.	20	20	20
1487	.	.	.	.	22	22	26
1488	.	.	.	.	42	42	47
1489	.	.	.	.	31	31	27
1490	.	21	.	.	25	25	27
1491	.	21	.	.	45	45	50
1492	.	.	.	.	.	.	.
1493	.	21	.	.	35	35	36
1494	.	.	.	.	18	18	18
1495	.	14	.	.	12	12	12
1496	.	.	.	.	42	42	40
1497	.	18	.	.	41	41	31
1498	.	29	.	.	26	26	26
1499	.	21	.	.	28	28	28
1500	.	12	.	.	14	14	14
1501	.	21	.	.	19	19	19
1502	.	.	.	.	29	29	26
1503	.	.	.	.	28	28	17
1504	.	.	.	.	14	14	17
1505	.	.	.	.	43	43	43
1506	.	29	.	.	28	28	29
1507	.	10	.	.	21	21	19
1508	.	17	.	.	30	30	32
1509	.	.	.	.	20	20	25
1510	.	.	.	.	30	30	30
1511	.	36	.	.	44	44	44
1512	.	.	.	.	24	24	24
1513	.	.	.	.	.	.	28
1514	.	29	.	.	37	37	29
1515	.	.	.	.	35	35	31
1516	.	.	.	.	12	12	11
1517	.	.	.	.	26	26	22
1518	.	.	.	.	32	32	28
1519	.	29	.	.	40	40	37
1520	.	7	.	.	35	35	23
1521	.	16	.	.	.	.	23
1522	.	29	.	.	5	5	24
1523	.	1	.	.	-5	-5	17
1524	.	.	.	.	.	.	14
1525	.	.	.	.	21	21	20

INTERPOLATION : 'V' FROM AVIGNON

```
APPENDIX    -    GRAPE HARVEST DATES
ORIGINAL AND INTERPOLATED SERIES
```

CORRELATION AND DIFFERENCE OF MEANS IN THE COMMON YEARS

	ALBENGA	BEAUNE	DIJON	MONCALIERI	AVIGNON	W.-EUROPE
ALBENGA	-	22 0.677 -3.46	18 0.579 1.50	26 0.230 3.27	1 -	-
BEAUNE	22 0.677 3.46	-	30 0.734 2.75	32 0.300 4.78	-	-
DIJON	18 0.579 -1.50	30 0.734 -2.75	-	42 0.379 -2.35	8 0.174 -8.63	41 0.825 0.36
MONACALIERI	26 0.230 -3.27	32 0.300 -4.78	42 0.379 2.35	-	5 -	23 0.354 10.22
AVIGNON	1 -	-	8 0.174 8.63	5 -	-	-
WEST-EUROP	-	-	41 0.825 -0.36	23 0.354 -10.22	-	-

```
1ST LINE : NUMBER OF COMMON VALUES, COEFFICIENT OF CORRELATION
2ND LINE : DIFFERENCE OF MEANS : TOP-SERIES MINUS LEFT-SERIES
```

References

Acknowledgments are due to Prof. John Post, Boston, for reading the manuscript and for making helpful suggestions, to Andreas Lauterburg and Hannes Schüle for providing invaluable help in programming and to Dr. P. Alexandre, Embourg (Belgium), for sending additional data.

Alexandre, P. 1986: Le Climat en Europe au Moyen-Age. Contribution à l'histoire des variations climatiques de 1000 à 1425, d'àpres les sources narratives de l'Europe occidentale. Paris (Ed. de l'Ecole des Hautes Etudes en Sciences Sociales).

Amberg, B. 1890, 1892, 1897: Beiträge zur Chronik der Witterung und verwandter Naturerscheinungen mit besonderer Rücksicht auf das Gebiet der Reuss und der angrenzenden Gebiete der Aare und des Rheines. In: Jahresber. Höh. Lehranst. Luzern.

Angot, A. 1885: Etude sur les Vendanges en France. Annales du Bureau Central Météorologique de France, Année 1883.

Annales, 1861: Annales Basilienses et Colmarienses ed. by Ph. Jaffe. Monumenta Germaniae Historica, scriptores, 17, 189-231. (quoted by Alexandre, 1986).

Becker, N. 1969: Beitrag zum Menge-Güte-Problem im deutschen Weinbau. In: Weinwissenschaft, 24, 172-190.

Bell, W., Ogilvie, A.E. 1978: Weather compilations as a source of data for the reconstruction of European climate during the medieval period. In: Climatic Change 1/4, 331-348.

Burga, C.A. 1985: Paläoklimatische Auswertung von Bündner Naturchroniken. In: Geographica Helvetica 40/4, 196-204.

Buszello, H. 1982: "Wohlfeile" und "Theurung" am Oberrhein 1340-1525 im Spiegel zeitgenössischer erzählender Quellen. In: Bauer Reich und Reformation. Festschrift für Günther Franz zum 80. Geburtstag, hg. von P. Blickle, 18-42, Stuttgart.

Dansgaard, W. 1984: Past climates and their relevance to the future. In: Flohn, Fantechi, 208-248.

Dubois, H. 1976: Les foires de Chalon et le commerce dans la vallée de la Saone à la fin du Moyen-age (vers 1280 - vers 1430). Paris.

Flohn, H., Fantechi, R. (ed.) 1984: The Climate of Europe: Past, Present and Future. Natural and Man-Induced Climatic Changes: A European Perspective. Dordrecht (Reidel).

Flohn, H. 1984: Ice-free Arctic and glaciated Antarctic. In: Flohn, Fantechi, 248-268.

Flohn, H. 1985: A critical assessment of proxy data for climatic reconstruction. In: The Climatic Scene, ed. by M.J. Tooley & G.M. Sheail, 93-103, London (Allen & Unwin).

Holzhauser, H.P. 1984: Zur Geschichte der Aletschgletscher und des Fieschergletschers. Physische Geographie, 13. Zürich (Geographisches Institut).

Hughes, M.K., Kelly, P.W., Pilcher, J.R., La Marche Jr, C.V. (eds): Climate from tree rings. Cambridge (Cambridge Univ. Press; quoted by Flohn, 1985).

Ingram, M., Underhill, D., Farmer G. 1981: The use of documentary sources for the study of past climates. In: Climate and History (Hrsg. T.M.L. Wigley et al., Cambridge), 180-213.

Koch, L. 1945: Meddelser om Gronland, 130, Nr. 3 (quoted in Dansgaard, 1984).

Lamb, H. 1982: Climate, history and the modern world. London (Methuen).

Lamb, H. 1984: Climate in the Last Thousand Years: Natural Climatic Fluctuation and Change. In: Flohn, Fantechi, 25-64.

Lavalle, M.J. 1855: Histoire et Statistique de la vigne et des grands vins de la Côte d'Or. Avec le concours de J. Garnier. Dijon (Picard).

Legrand, J.P. 1979a: Les fluctuations Météorologiques exceptionnelles durant les Saisons Printanières et Estivales depuis le Moyen-Age. In: La Météorologie, 6th series, 16, 167-181; 18, 131-141.

Legrand, J.P. 1979b: L'expression de la vigne au travers du climat depuis le moyen-age. Revue Française d'Oenologie 75, 23-52.

Le Mené, M. 1982: Les campagnes Angevines a la fin du Moyen-Age (vers 1350 - vers 1530). Etude économique, Nantes.

Le Roy Ladurie, E., Baulant, M. 1980: Grape harvests from the fifteenth through the nineteenth centuries. In: J. of Interdisciplinary History 10/4, 839-849.

Le Roy Ladurie, E. 1971: Times of Feast, Times of Famine. A history of climate since the year 1000. London (Allen & Unwin). (Translation of the 1967 French edition: L'histoire du climat depuis l'an mil. Paris).

Mazzei, M.: Le variazioni del clima nella piana di Albenga in eta' moderna. Thesis at the University of Genova, Faculty of Letters. (Source and data from Alexandre, 1986).

McGovern, T.H. 1981: Economics of extinction in Norse Greenland. In: Wigley et. al., Climate and History. Studies in past climates and their impact on man, Cambridge, 403-433.

Pavese, M.P., Gregori, G.P. 1985: An analysis of six centuries (XII through XVII century A.D.) of climatic records from the Upper Po Valley. In: W. Schröder (ed.), Historical events and people in geosciences, Frankfurt, 185-220.

Pfister, Ch. 1984: Klimageschichte der Schweiz 1525-1860. Das Klima der Schweiz von 1525-1860 und seine Bedeutung in der Geschichte von Bevölkerung und Landwirtschaft. Vol 1. Bern (Haupt).

Pfister, Ch. 1985a: CLIMHIST - a weather data bank for Central Europe 1525 to 1863. May be ordered from METEOTEST, Hallerstrasse 50, CH-3012 Berne.

Pfister, Ch. 1985b: Snow cover, snow lines and glaciers in Central Europe since the 16th century. In: The Climatic Scene, ed. by M.J. Tooley and G.M. Sheail, London, 154-174.

Pfister, Ch. 1985c: Veränderungen der Sommerwitterung im südlichen Mitteleuropa von 1270-1400 als Auftakt zum Gletscherhochstand der Neuzeit. In: Geographica Helvetica 40/4, 186-195.

Rotelli, C. 1973: Una campagna medievale. Storia agraria del Piemonte fra il 1250 e il 1450. Torino (Einaudi).

Röthlisberger, F. 1976: Gletscher- und Klimaschwankungen im Raum Zermatt, Ferpècle und Arolla. In: 8000 Jahre Walliser Gletschergeschichte. Ein Beitrag zur Erforschung des Klimaverlaufs in der Nacheiszeit. Die Alpen 52/3-4, 58-152.

Scherer, G.S. 1874: Kleine Toggenburger Chronik, St. Gallen.

Schmitz, H.J. 1968: Faktoren der Preisbildung für Getreide und Wein in der Zeit von 800 bis 1350. Stuttgart (Fischer).

Schweingruber, F.H., Fritts, H.C., Bräker, U., Drew, G., Schär, E. 1978: The X-ray technique as applied to dendroclimatology. In: Tree-Ring Bulletin 38, 61-91.

Schweingruber, F.H., Bräker, O.U., Schär, E. 1979: Dendroclimatic studies on conifers from central Europe and Great Britain. In: Boreas 8, 427-452.

U.S. Department of Energy 1985: A climate data-bank for northern hemisphere land areas 1851-1980. TRO No 17.

WMO 1986: Report of the International Conference on the Assessment of the Role of Carbon Dioxide and of Other Greenhouse Gases in Climate Variations and Associated Impacts. Villach, Austria, 9-15 October 1985, WMO No. 661.

NORWEGIAN SEA DEEP WATER VARIATIONS OVER THE LAST CLIMATIC CYCLE: PALEO-OCEANOGRAPHICAL IMPLICATIONS

J. C. Duplessy , L. Labeyrie
Centre des Faibles Radioactivités
Laboratoire mixte CNRS-CEA
F-91190 Gif sur Yvette, France

and

P.L. Blanc
C.E.A./I.P.S.N./D.P.T.
CEN/FAR
B.P. 6-92265 Fontenay aux Roses Cedex, France

1.   Introduction

The Norwegian Sea is one of the most critical areas of the modern world ocean because of the large volume of deep dense water which is formed there by intense surface cooling during winter. This deep water flows out over the Denmark-Faeroe strait and forms a major portion of North Atlantic Deep Water (NADW), the most important deep-water type in the world ocean, because it penetrates all oceanic basins until reaching the North Pacific (Reid and Lynn, 1971).

Cores from the ocean floor are made of material deposited at a rate of a few cm per thousand years. This material contains shells of planktonic foraminifera (animals that live in surface waters) and benthic foraminifera (that live on the bottom). The occurrence of the heavy oxygen isotope, ^{18}O, is an important parameter in deep sea core analysis, because the $^{18}O/^{16}O$ ratio of the carbonate formed by micro-organisms and deposited in marine sediments is dependent on both the temperature and isotopic composition of the ocean water at the time and place the micro-organisms lived (Emiliani, 1955; Duplessy, 1978). $^{18}O/^{16}O$ ratios are expressed as the relative deviation of the $^{18}O/^{16}O$ ratio in a sample from that in a standard. At any given time, the $^{18}O/^{16}O$ ratio of all ocean water must be the higher the more isotopically light (low $^{18}O/^{16}O$ ratio) water is removed from the ocean and deposited as glacier ice on the continent. In order to estimate past sea surface temperature from isotopic analysis, it is thus necessary to estimate independently the time-variations of the global ocean $^{18}O/^{16}O$ ratio. This is the goal of the present paper.

Micropaleontological and isotopic analysis of Norwegian Sea cores have shown that the conditions prevailing at the Norwegian Sea surface have not remained constant over the last climatic cycle and this sea has been covered with ice during most of the last ice age (Kellogg, 1980). As a consequence, the mechanism by which deep water is formed today was not active and the Norwegian Sea was not a source of deep water for the glacial ocean (Duplessy et al., 1975). These authors also demonstrated that the amplitude of the 150 kyr oxygen isotope record of benthic foraminifera was smaller than in the other oceanic basins and that the glacial $\partial^{18}O$ value was similar to that of benthic foraminifera from the deep Atlantic Ocean, indicating that both deep waters had similar hydrological characteristics. By using a model in which the Pacific Deep Water was assumed to have remained at a constant temperature, Duplessy et al. (1975) suggested that the temperature of the Norwegian Sea Deep Water could have been warmer during the last ice age than today.

However, the recent evidence that deep water formed in the Norwegian Sea after the last interglaciation during isotope substage 5d, about 107 kyr ago (Duplessy and Shackleton, 1985) and that the high latitude North Atlantic was a source of cold deep water during the whole glaciation (Duplessy et al, 1980; 1987; Boyle and Keigwin, 1982; Mix and Fairbanks, 1985) contradicts the hypothesis that Norwegian Sea Deep Water was noticeably warmer during the last ice age. It is thus necessary to re-assess the reconstruction of the evolution of the Norwegian Sea Deep Water during the last climatic cycle and to determine when it was cut off from the world ocean circulation.

2.   Strategy

The isotopic paleoceanography of the Norwegian Sea is made difficult for several reasons. First, the sedimentation rates vary considerably and all time intervals are not recorded in a single sediment core (Kellogg et al., 1978). It is thus necessary to analyze and compare several cores in order to generate a complete climatic record.

Second, a paleoclimatic reconstruction must rest on foraminiferal shells in place and displaced fauna is a severe problem. Holtedahl (1959) showed that an entire fauna has been displaced downslope in the late glacial in the southeastern Norwegian Sea. Streeter et al. (1982) have shown that the percentage of displaced specimens is high during glacial conditions and related this displacement to ice rafting, which can be easily recognized by the high percentage of quartz in the sediment. During the last glacial maximum, these authors showed that the sediment is barren from benthic (i.e. deep water) foraminifera in place, whereas the abundance of displaced benthic fauna may be high. Third, porcelaneous species, noticeably *Pyrgo* , are abundant in portions of the cores. However, repeated analysis of the isotopic composition of specimens of this genus have shown an unexplained dispersion with deviations up to 1 per mil toward either light or heavy isotope value (Duplessy et al., 1984). As *Pyrgo* was used by Duplessy et al. (1975) to generate the isotope record of core K-11, it may contribute to inaccuracies in this record. The most abundant hyaline benthic species are *Cibicides wuellerstorfi* and *Oridorsalis tener* in the Holocene sediment (Belanger and Streeter, 1980). Streeter et al. (1982) showed that *Cibicides* is dominant during Holocene, isotope stages 4 and 5 and is absent during isotope stages 2, 6 and most of stage 3. By contrast, *Oridorsalis* becomes dominant when *Cibicides* is rare or absent, noticeably during glacial conditions. We therefore used only these two species in order to generate a benthic $\partial^{18}O$ record extending to the last interglaciation.

In this paper, we present analyses of five sediment cores from the Norwegian Sea (Fig. 1). Since none of them yielded a complete record of the conditions prevailing during the last climatic cycle, we developed a detailed stratigraphic framework in order to express the $\partial^{18}O$ variations relative to a common time scale and we generated a stacked benthic isotope record describing the hydrological evolution of the Norwegian Sea Deep Water. We then present an example illustrating how the difference between this record and that of either benthic or planktonic foraminifera from the major oceanic basins may be interpreted in terms of temperature variations for the ocean water in the past.

3.   Isotopic Calibration of the Benthic Species

Since benthic fauna vary in response to environmental and climatic changes, generally any single benthic foraminiferal species is not abundant enough for analysis throughout the length of a deep sea core. In order to get a complete isotopic record, one reference species is chosen and the other species are compared to that reference species. Following the recommendation of Duplessy et al. (1984), we adjusted in this paper $\partial^{18}O$ values to *Uvigerina*, because this adjustment provides the oxygen isotopic composition of calcite in equilibrium with sea water at the time of deposition.

The calibration of *Cibicides* versus *Uvigerina* for $\partial^{18}O$ is now well established by thousands of analyses (Duplessy et al., 1970; 1984; Shackleton and Opdyke, 1973; Shackleton and Cita, 1979; Blanc and Duplessy, 1982) and we applied the classical adjustment of +0.64. By contrast, *Oridorsalis* tener was very poorly calibrated for both oxygen and carbon. We made this calibration by comparing the $\partial^{18}O$ and $\partial^{13}C$ values of *O. tener* and *C. wuellerstorfi* during isotope stages 1, 4, and 5, when both species were abundant and clearly coexisted. These data (Table 1 in Appendix) show that it is reasonable to adjust the $\partial^{18}O$ of

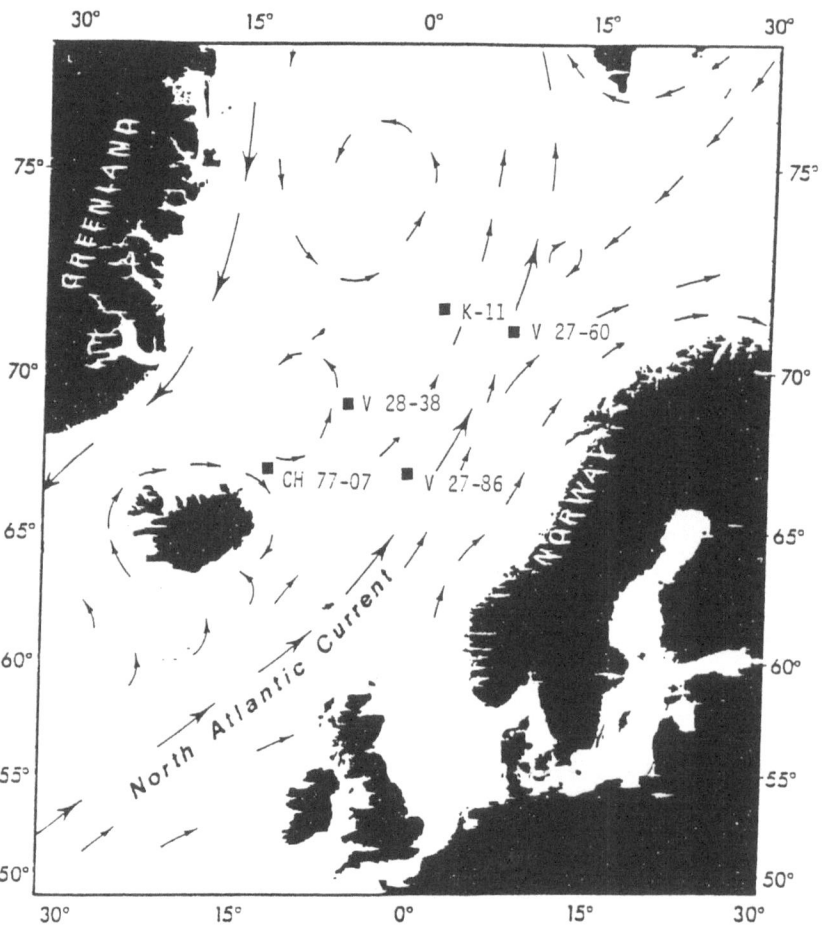

Figure 1 : Core location.

CH 77-07	66°36' N	10°31' W	1487 m
K-11	71°47' N	1°36' W	2900 m
V 27-60	72°11' N	8°35' E	2525 m
V 27-86	66°36' N	1°07' E	2900 m
V 28-38	69°23' N	4°24' W	3411 m

O. tener by adding +0.36 to the measured values. Conversely, the $\partial^{13}C$ difference between *O. tener* and *C. wuellerstorfi* is not constant with time and our data indicate that it is not possible to estimate past $\partial^{13}C$ of ΣCO_2 dissolved in sea water by analyzing *O.tener*. Such a difficulty is not unusual and has already been observed for other benthic species, such as *Uvigerina* (Zahn et al., 1986). We therefore did not attempt any adjustment of the measured $\partial^{13}C$ values of *O.tener* and used only the $\partial^{13}C$ values of *C.wuellerstorfi* in order to estimate the carbon isotopic composition of ΣCO_2 dissolved in the Norwegian Sea deep water.

4.   Displaced Foraminiferal Shells in Glacial Sediment

Evidence for the presence of displaced shells, probably transported with ice rafted material, is abundant (Holtedahl, 1959; Belanger and Streeter, 1980; Streeter et al., 1982). We observed that in large sediment samples from core K-11, small amounts of *C. wuellerstorfi* shells may be found in glacial sediment and the question arises whether these specimens are in place or displaced. We therefore made a detailed comparison of the $\partial^{18}O$ and $\partial^{13}C$ variations of *C. wuellerstorfi* and *O. tener* in this core. Fig. 2 shows that whereas *O. tener* exhibits isotopic variations roughly correlated with those of *N. pachyderma* , *C. wuellerstorfi* had a constant isotopic composition during the whole isotope stages 2 and 3, characteristics of interglacial climate. By contrast, during isotope stages 1, 4 and 5, when both species are abundant, they provide a similar record. We thus believe that the few *C. wuellerstorfi* shells found in glacial sediment are displaced specimens.

We therefore analyzed *C. wuellerstorfi* in the other cores only during stages 1, 4, and 5, when this species is abundant. This strategy minimizes the perturbation effects of both bioturbation and transport (Bard et al., 1987).

5.   The Planktonic and Benthic Isotope Record

The oxygen and carbon isotope ratios of *N. pachyderma, O.tener* and *C.wuellerstorfi* in the five cores are reported in Table 1. *N. pachyderma* left coiling has been used to derive the planktonic record, because it is adapted to very low temperature and is the most abundant planktonic (i.e. surface water) foraminifer in the Norwegian Sea. This species develops mostly during spring to summer. Its $\partial^{18}O$ is in isotopic equilibrium for the hydrological conditions corresponding to a mean depth habitat of 80-100 m in the Norwegian Sea (Kellogg et al., 1978).

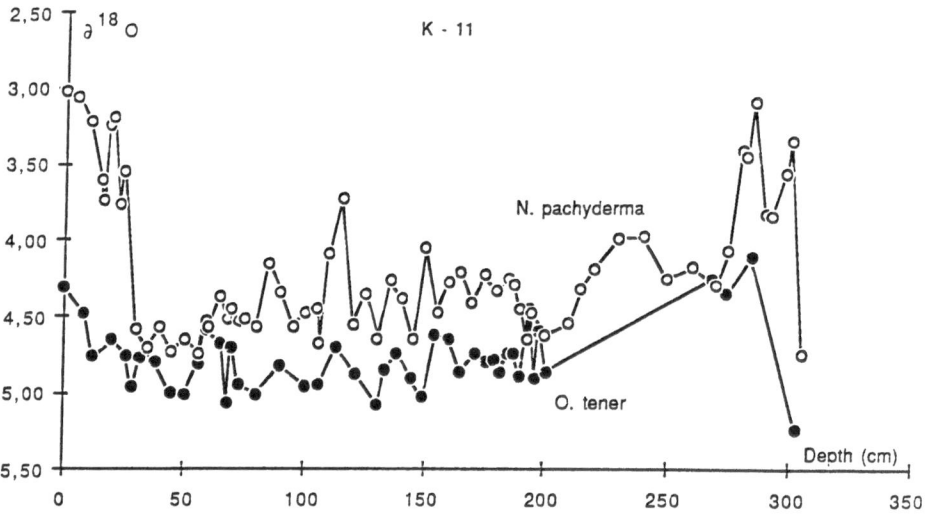

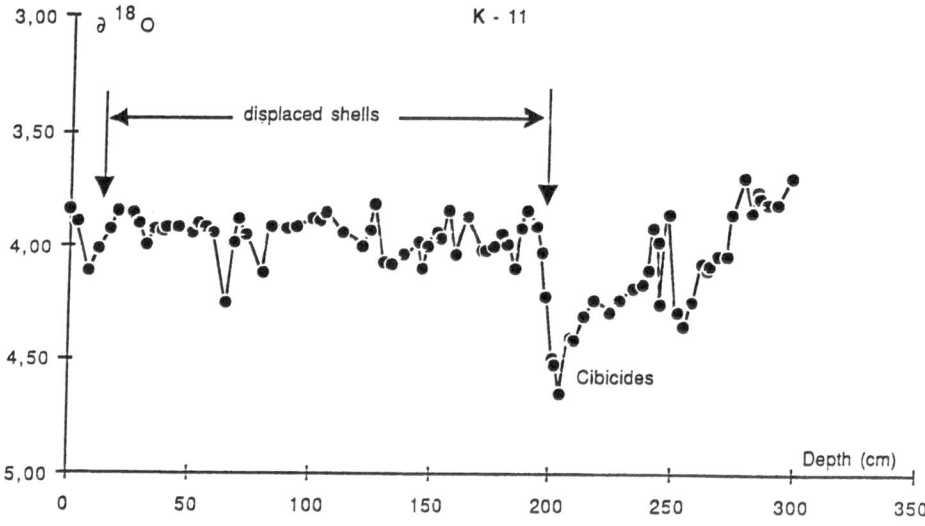

Figure 2 : Oxygen isotope record of *N. pachyderma, O. tener* (upper curves) and *C. wuellerstorfi* (lower curve) in core K-11. Constant isotopic ratios of the rare *C. wuellerstorfi* shells found in glacial sediment are interpreted as an indication that these shells have been transported and are not indicative of local conditions at the time of deposition of the sediment. *N. pachyderma* lives in surface water. *O. tener* and *C. wuellerstorfi* live at depth on the bottom.

N. pachyderma occurs throughout nearly the whole of the last climatic cycle (Fig.3-7). The $\partial^{18}O$ benthic record has been established by analyzing monospecific samples. The analyses are plotted (Fig. 3-7) with correction of +0.64 for *C. wuellerstorfi* and +0.36 for *O.tener*, to take account of their respective departure from isotopic equilibrium. When several measurements were made at the same level, only their mean value was plotted.

The $\partial^{18}O$ records of *N. pachyderma* in Norwegian Sea sediment cores reflect the world-wide $\partial^{18}O$ signal (Shackleton and Opdyke, 1973) modified somewhat by temperature and salinity effects peculiar to this area (Kellogg et al., 1978). The records of cores K-11, V 27-60, V 27-86 and V 28-38 extend until the boundary between isotope stages 6 and 5 and show a great similarity. Their amplitude is larger than 2 per mil and they all exhibit a sharp peak corresponding to isotope substage 5e. By contrast, the benthic record exhibits a much smaller amplitude, which is close to 1 per mil. The most striking difference between the planktonic and the benthic records is observed at the transition between substages 5e and 5d, which marks the beginning of glaciation: The mean amplitude of $\partial^{18}O$ change between the peak of 5e and the maximum value of 5d is 0.75 per mil for the benthics whereas it is 1.6 per mil for the planktonic foraminifera. This difference between the planktonic and the benthic signal cannot be explained by bioturbation, because the benthic foraminifera are abundant during the whole stage 5. It therefore implies that the summer surface waters and the bottom waters evolved in very different ways during the beginning of glacial conditions.

The detailed records of core V 28-38 and V 27-60 also show that the $\partial^{18}O$ increase of *N. pachyderma* preceeds that of the benthics. This suggests that a cooling of surface waters by 3-4°C occurred during isotope substage 5e, before the development of a significant amount of ice over the continents. This cooling phase is followed by a small isotopically light peak, which does not coincides with isotopic substage 5c, as defined in the benthic record, but preceeds it by more than 10 cm in all the records. This illustrates that the details of the isotopic record of *N. pachyderma* in the high latitudes are not easily correlated with those of lower latitudes.

Figures 3-7 also display the $\partial^{13}C$ variations of *N. pachyderma* and *C. wuellerstorfi* plotted against depth in the five cores. A common feature to all the *C. wuellerstorfi* records is the presence of high $\partial^{13}C$ values during the whole of stage 5 and stage 4. These values are more positive than those measured in benthic records from the North Atlantic (Shackleton, 1977; Duplessy, 1982; Sarnthein et al., 1984; Mix and Fairbanks, 1985; Zahn et al., 1986) or the other oceanic basins (Shackleton et al., 1984) and are indicative of recent contact with the atmosphere (Duplessy and Shackleton, 1985). Gas exchange with the atmosphere was possible at that time, since Belanger (1982) observed the presence of coccoliths in Norwegian Sea bottom sediments throughout stage 5 and into stage 4. The presence of coccoliths implies that summer

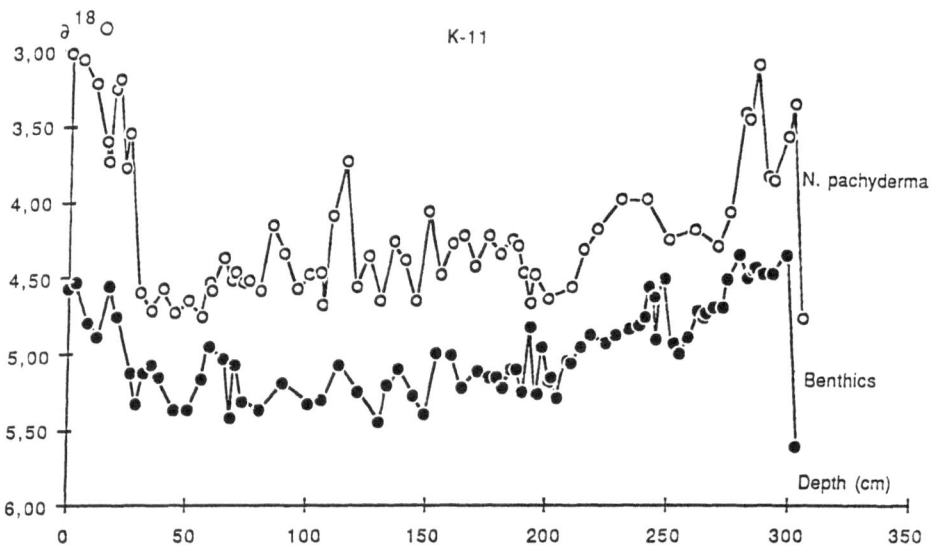

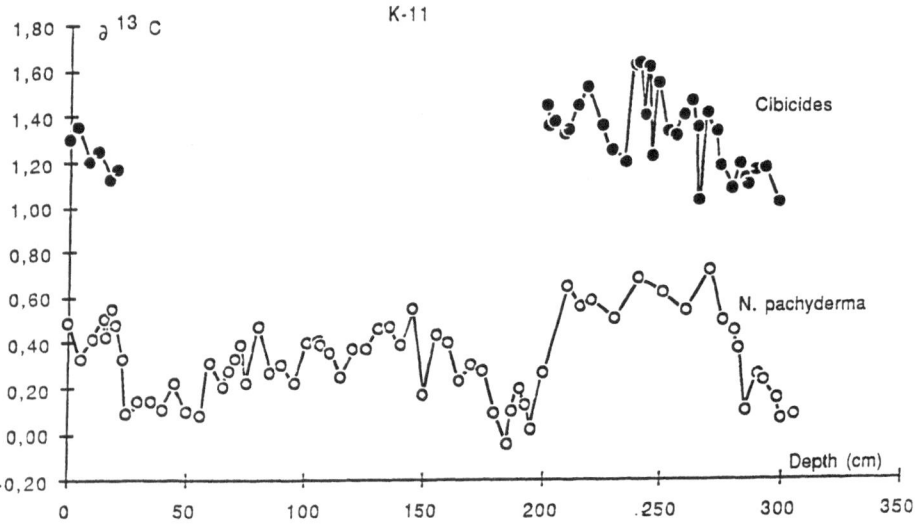

Figure 3 :   Oxygen and Carbon isotope record of *N. pachyderma* ,
Carbon isotope record of *C. wuellerstorfi* and Oxygen isotope
record of benthic calcite in core K-11. *N. pachyderma* lives in
surface water. *C. wuellerstorfi* lives at depth on the bottom.

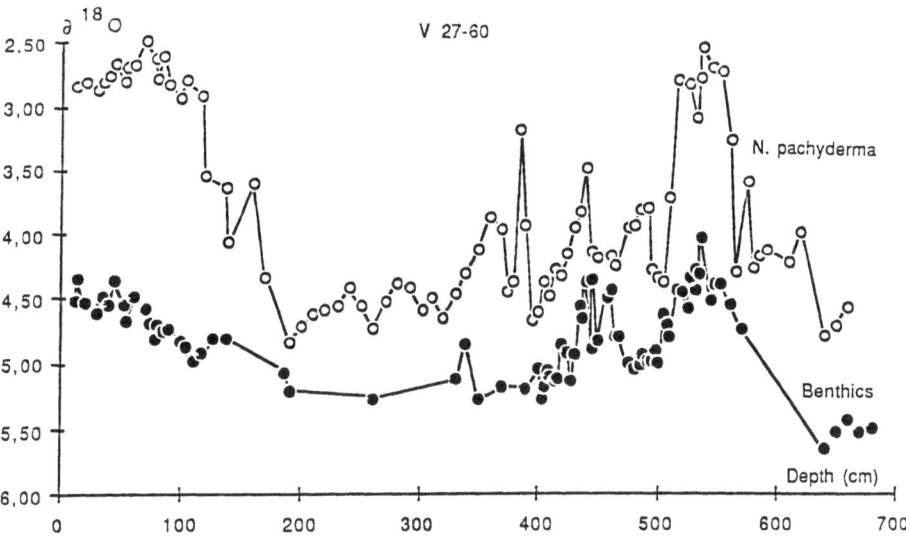

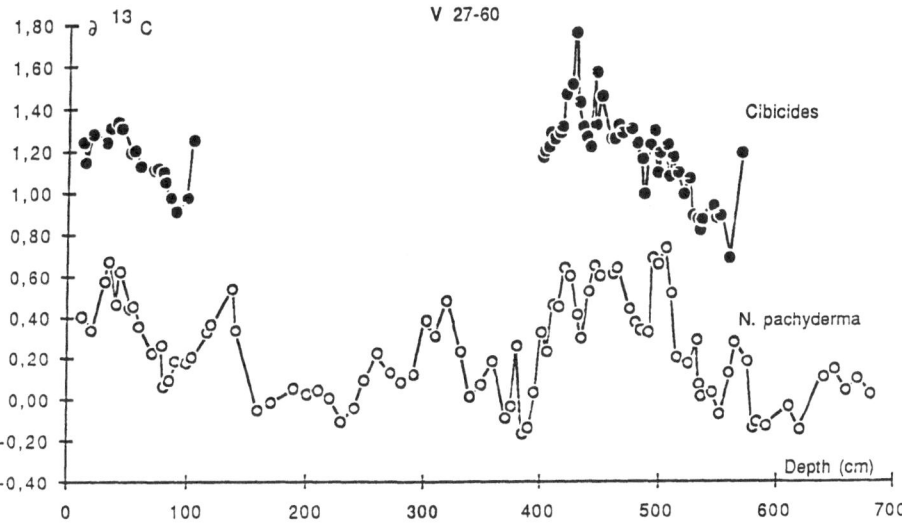

Figure 4 : Oxygen and Carbon isotope record of *N. pachyderma* ,
Carbon isotope record of *C. wuellerstorfi* and Oxygen isotope
record of benthic calcite in core V 27-60. *N. pachyderma* lives
in surface water.*C. wuellerstorfi* lives at depth on the bottom.

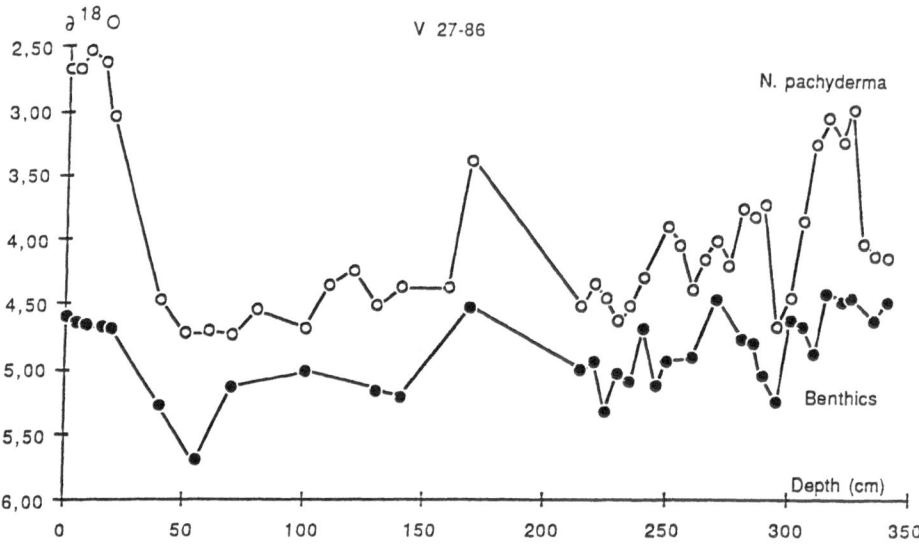

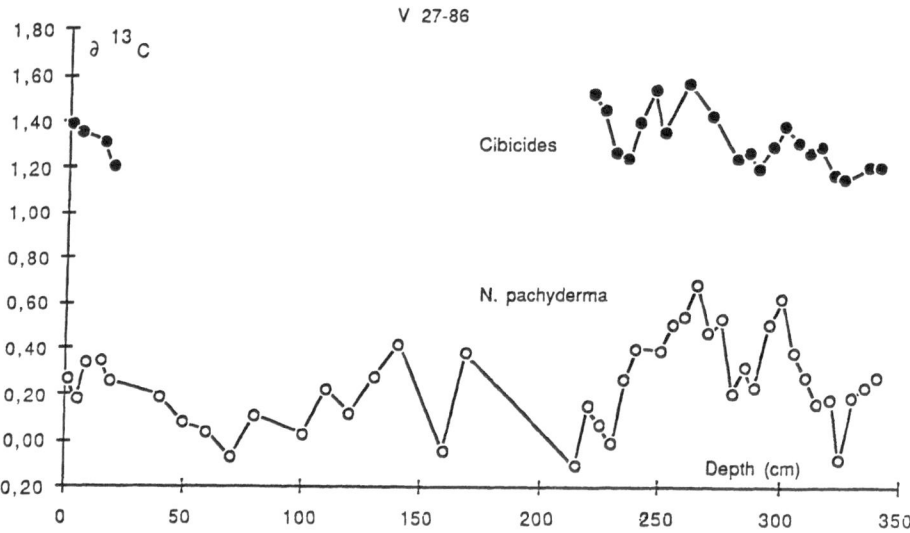

Figure 5 : Oxygen and Carbon isotope record of *N. pachyderma* ,
Carbon isotope record of *C. wuellerstorfi* and Oxygen isotope
record of benthic calcite in core V 27-86. *N. pachyderma* lives
in surface water.*C. wuellerstorfi* lives at depth on the bottom.

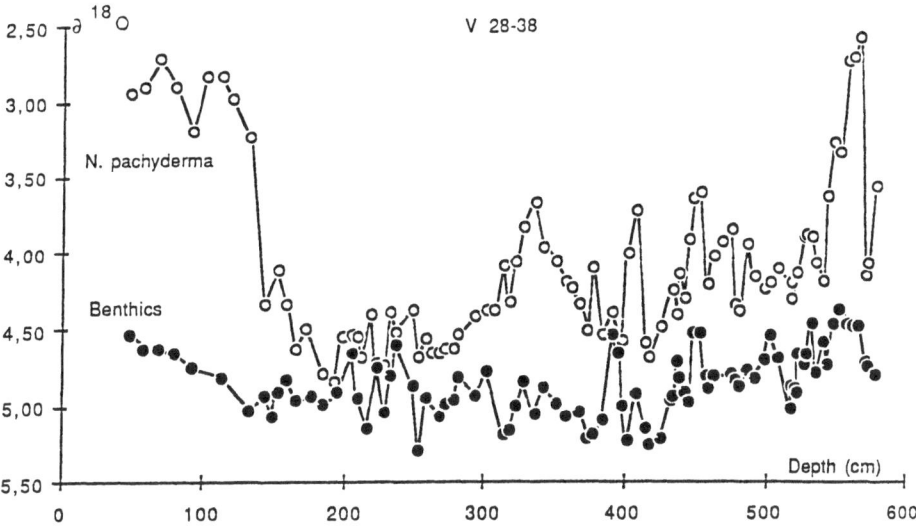

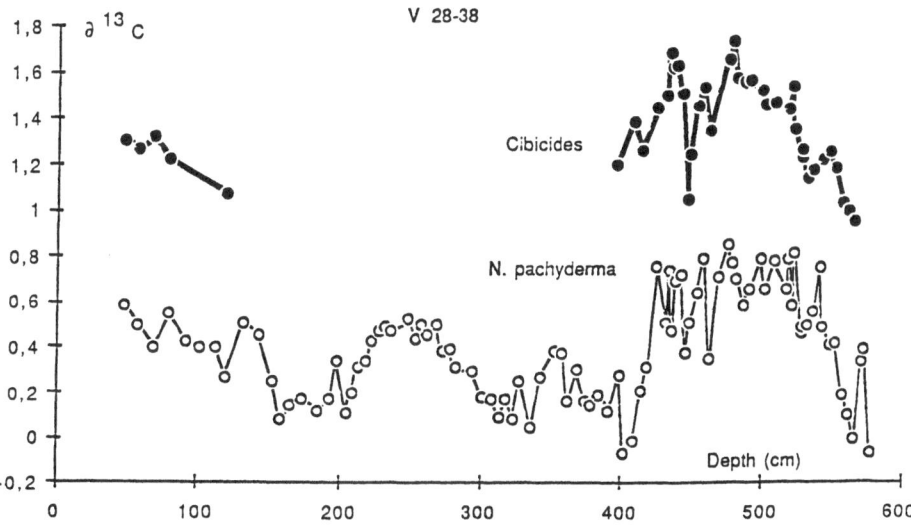

Figure 6 : Oxygen and Carbon isotope record of *N. pachyderma* ,
Carbon isotope record of *C. wuellerstorfi* and Oxygen isotope
record of benthic calcite in core V 28-38. *N. pachyderma* lives
in surface water. *C. wuellerstorfi* lives at depth on the bottom.

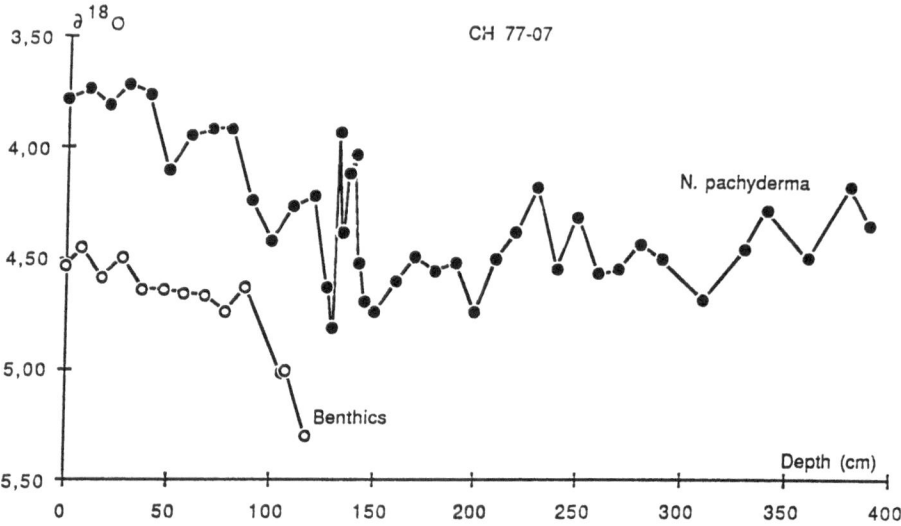

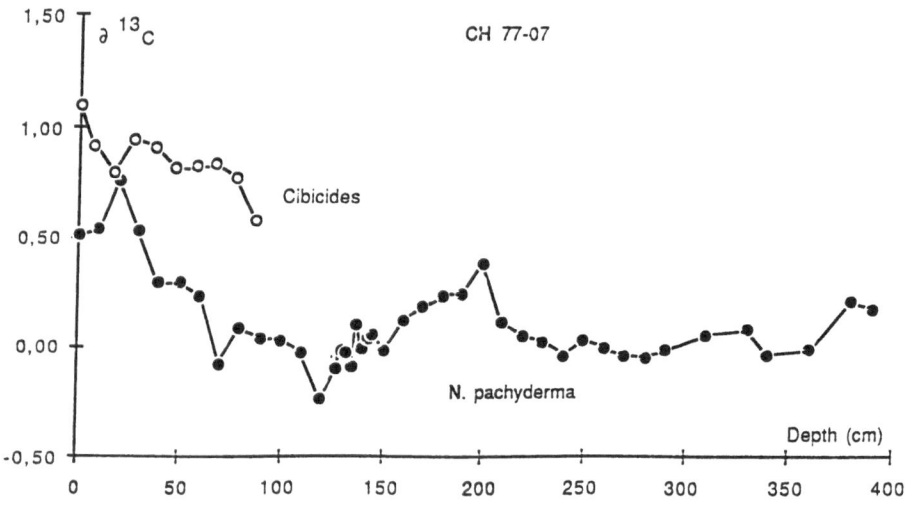

Figure 7: Oxygen and Carbon isotope record of *N. pachyderma* , Carbon isotope record of *C. wuellerstorfi* and Oxygen isotope record of benthic calcite in core CH 77-07. *N. pachyderma* lives in surface water. *C. wuellerstorfi* lives at depth on the bottom.

temperatures in the photic zone at that time were warmer than 2-3°C, as coccospheres do not live today at lower temperatures. We thus conclude that the Norwegian Sea was free of ice, at least during summer and that it was an active site of deep water formation during this time (from about 128 to 62 Kyr B.P.). The active convection in the Norwegian Sea is also supported by the high degree of correlation between the $\partial^{13}C$ records of the planktonic and benthic foraminifera (r = 0.80 for core V 28-38, 0.73 for core V 27-60, 0.62 for core K-11 and 0.40 for core V 27-86, which has a poor resolution).

The abundance of *C. wuellerstorfi* sharply decreases by the end of stage 4 (in cores V 27-60, K-11, V 27-86) or at the early beginning of stage 3 (in core V 28-38), whereas *O.tener* is still present. Under modern conditions, *C. wuellerstorfi* is the most abundant hyaline species in Norwegian Sea sediments at a water depth between 1250 and 2900 m (Belanger and Streeter, 1980; Mackensen et al., 1985) and *O. tener* becomes dominant below 2900 m. From our data alone, we cannot tell whether *C. wuellerstorfi* was drastically reduced in abundance during glacial conditions or whether the faunal boundary which is at present at 2900 m migrated upward. However, we believe that this species probably disappeared almost entirely from the Norwegian Sea during the remaining of the glaciation, because its absence in glacial sediments has also been observed in shallower cores (Streeter et al, 1982; Sejrup et al., 1984).

Streeter et al. (1982) interpreted the change in dominance from *C. wuellerstorfi* to I *O.tener* as an abrupt shift to an ice-covered Norwegian Sea. Deep water formation would not be possible in the presence of a permanent ice cover and we believe that the benthic fauna shift coincides with the development of hydrologic conditions which did not permit deep water formation. This hypothesis is supported by the very negative $\partial^{13}C$ values measured in *O. tener* from glacial sediments, that are best explained by the presence of a deep water-mass poorly oxygenated at that time. With this hypothesis, the re-appearance of *C. wuellerstorfi* in all the cores by the end of the second step of the deglaciation (Termination IB, Duplessy et al., 1981). dates the renewal of the deep water ventilation. Sejrup et al. (1984) reported that *C. wuellerstorfi* reinhabited the Norwegian Sea at shallower depth within the first part of the deglaciation (Termination IA), about 5,000 years earlier. This suggests that the convection started during the middle of the deglaciation, but that the water mixing became strong enough to renew the whole Norwegian Sea deep water mass only at the end of the melting phase of the continental ice sheet.

6.   Absolute Chronology

We can identify several isotopic datum levels in the $\partial^{18}O$ records of N. pachyderma and of the benthic species, which are widely recognized and dated outside the Norwegian Sea: Termination I, which has been dated by 14C Accelerator mass-spectrometry (Duplessy et al., 1986; Bard et al.,1987). The 4/5a and 5e/6 boundaries can be clearly identified in all the cores and their age has been assigned within the framework of the astronomical theory of the Pleistocene climate (Imbrie et al., 1984).

Other isotopic boundaries are not easily recognized, noticeably within stages 3 and 5. We refined the correlation between the 5 cores through their $\partial^{13}C$ records of N. pachyderma, which present a close similarity over the last 140,000 years in the high latitude of the North Atlantic and in the Norwegian Sea (Labeyrie and Duplessy, 1985). A final adjustment has been made by matching the events recognized in the $\partial^{18}O$ benthic records, because the Norwegian Sea Deep Water must have had rather uniform properties within this single basin.

The chronology applied to this stratigraphic framework is obtained by piecing that of Duplessy et al. (1986) for the last 25,000 years with that of Paterne et al. (1986) for isotope stage 3 and with the SPECMAP time-scale for stages 4 and 5 (Imbrie et al., 1984). The estimated ages of the levels taken as reference in each core are reported in Table 2.

The $\partial^{13}C$ record of N. pachyderma exhibits isotopically light peaks, which roughly correlate with the oxygen isotope stages 2 and 4. We estimated the age of the 4 major transitions, because they may be used as additional stratigraphic markers. Results (Table 3) show that whereas the limits of the peak associated with isotopic stage 2 are approximately in phase with the oxygen isotope record, those associated with isotopic stage 4 lag the $\partial^{18}O$ record by about 5,000 years.

7. Stacked Benthic Oxygen Isotope Record from the Norwegian Sea

For each core, we used the normalized $\partial^{18}O$ values of the benthic foraminifera and the reference levels reported in Table 2. We calculated the age of each level in the core by linear interpolation between the upper and lower reference levels. We put together the data of the five cores in order to obtain a matrix with two columns, one for the age and the other for the benthic calcite $\partial^{18}O$ value. We then estimated the $\partial^{18}O$ value each 1/3 kyear by cubic spline interpolation and filtered the resulting record by a 15 point least square quadratic filter (Savitzky and Golay, 1964). Finally, we calculated from the filtered record the benthic calcite $\partial^{18}O$ value each thousand years. These $\partial^{18}O$ values are plotted against time in Fig. 8.

The major trends of the Norwegian Sea $\partial^{18}O$ benthic record can be easily recognized, in particular the low amplitude of the 5e/5d transition, which is much smaller than in the Atlantic, Indian and Pacific oceans. The oscillations corresponding to isotopic substages 5a-5d are well marked, but the $\partial^{18}O$ values corresponding to isotopic

Table 2: Age assigned to reference levels.

Core CH 77-07		Core K-11		Core V27-60		Core V27-86		Core V28-38	
Depth (cm)	Age (kyr)	Depth (cm)	Age (kyr)	Depth (cm)	Age (kyr)	Depth (cm)	Age (kyr)	Depth (cm)	Age (kyr)
0	0	0	5	0	0	0	3	0	0
100	9,8	15	10,5	70	6	9	6	69	6
		20	13	140	10,5	30	10,5	112	10,5
		30	15	160	13	50	15	164	15
		59	25	190	15	80	25	204	25
		90	32,7	240	25	120	37	222	32,7
		100	33,5	280	37	130	40,7	228	33,5
		113	37	320	40,7	169	48,8	236	37
		130	40,7	360	50,3	215	62,3	253	40,7
		154	44,8	375	56,1	227,5	65	302	45
		171	50,3	385	60,3	243	80	316	47,9
		186	51,8	403	65	260	87	328	48,8
		190	56,1	405	68	270	99	342	51,8
		193	57,4	420	71	280	112	376	56,1
		198	60,3	440	80	300	118	391	57,4
		204	65	450	87	325	122	408	59,3
		218	71	461	99	340	127	414	65
		242	80	475	108			436	71
		255	87	485	112			450	80
		270	99	500	116			481	87
		272	100,5	515	119			502	99
		274	117	536	122			517	107
		285	122	560	127			528	112
		300	127	640	135			532	113,3
		303	135					541	118
								555	122
								578	128

Table 3: Age of the 4 major transitions in the carbon record N. pachyderma in Norwegian Sea sediment cores.

Transition	K-11	V27-60	V27-86	V28-38	mean Age (kyr)	St. dev. (kyr)
1/2	13,6	11,8	15	14	13,6	1,3
2/3	25	28	25	26,7	26,2	1,5
3/4	50,7	59	56	58,4	56,0	3,8
4/5	61,9	64	70	66,1	65,5	3,5

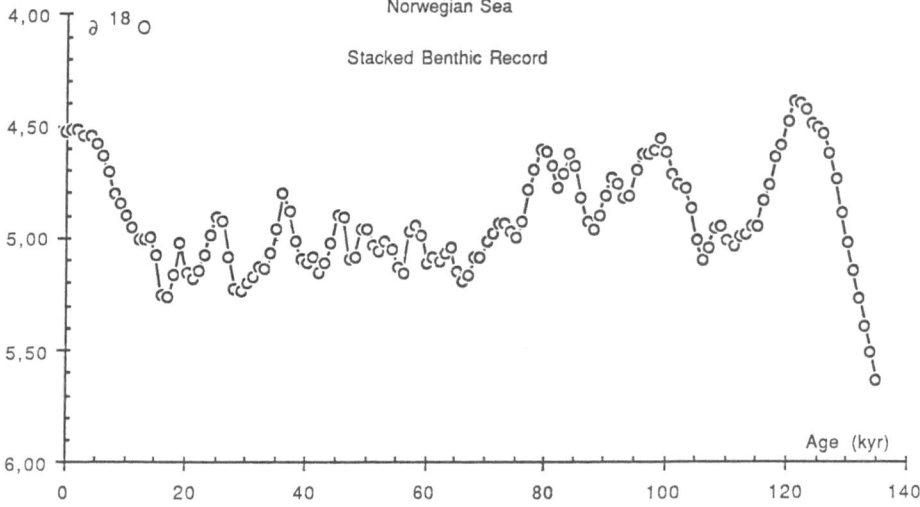

Figure 8 : Oxygen isotope record of benthic calcite in Norwegian Sea sediment during the last climatic cycle calculated by stacking the benthic records of the 5 Norwegian Sea cores.

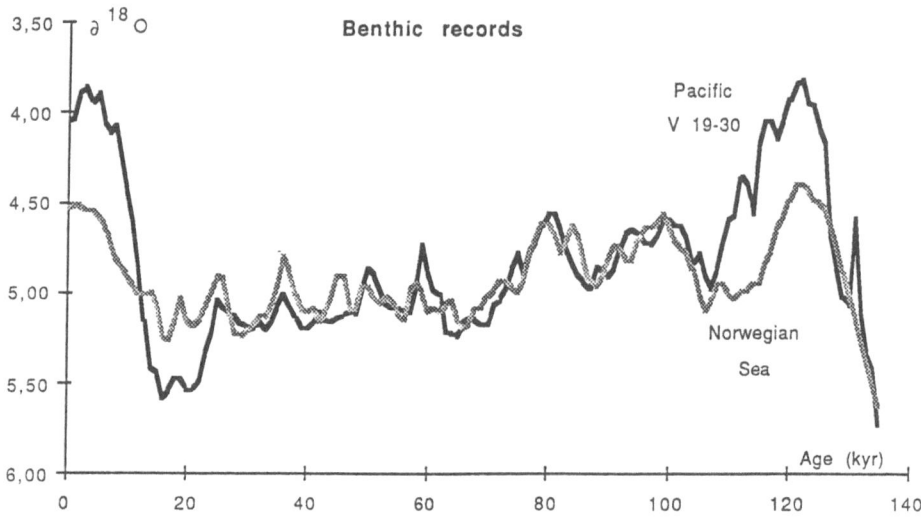

Figure 9 : Comparison of the stacked oxygen isotope record of benthic foraminifera from the Norwegian Sea with that of Pacific core V 19-30 analyzed by N. J. Shackleton at Cambridge.

substages 5a and 5c are only 0.1 per mil heavier than those measured during full interglacial conditions (isotope substage 5e). Since the *Cibicides* $\partial^{13}C$ values indicate that the Norwegian Sea was a sink for cold surface water during isotope stages 5 and 4, the Norwegian Sea Deep Water remained at a low temperature during this whole period. The proximity to sea-ice cover, which is well documented for isotopic substage 5a (Kellogg, 1980), implies that the sea surface temperature was cooler than today. These low temperature are confirmed by the disappearance of subpolar species during substages 5a-5d (Kellogg, 1976). As a consequence, the Norwegian Sea Deep Water, which was formed from this cold surface water, could not be warmer than today. Its present temperature is very low (-1°C) and a lower limit for the temperature of this water mass at depth is -1.7°C, taking into account the adiabatic warming that surface waters experience during sinking. As a first approximation, we can therefore assume that the Norwegian Sea Deep Water did not experience any significant temperature variation during the stages 5 and 4. In this case, the benthic foraminiferal $\partial^{18}O$ record has only registered the variations of the world ocean $\partial^{18}O$, resulting from the growth and decay of the northern hemisphere ice sheets.This hypothesis is supported by the sea level record, which shows variations closely similar to the Norwegian Sea stacked benthic record (Chappell and Shackleton, 1986; Dodge et al., 1983).

During the glacial stages 2 and 3, *C. wuellerstorfi* is absent and the same rationale cannot be applied since we cannot interpret the $\partial^{13}C$ variations of *O. tener* in terms of those of the total dissolved CO_2. We shall thus compare the stacked benthic Norwegian Sea record with that of Pacific Core V19-30, which has been closely tied to sea level variations (Chappell and Shackleton, 1986). Fig. 9 shows that the two isotope records are highly correlated during stages 3 to 5d (i.e. from 28,000 to 115,000 yr B.P.). Since the isotopic record of core V19-30 during this time interval parallels the sea level record, we conclude that the benthic Norwegian Sea record reflects there only sea water $\partial^{18}O$ variations and therefore the global ice volume changes. The sharp $\partial^{18}O$ increase corresponding to isotope stage 2 is not observed in the Nor- wegian Sea record. Three explanations are possible for this observation: first, the Norwegian Sea record of stage 2 may be considered as a global sea water $\partial^{18}O$ record. In that case, the Pacific $\partial^{18}O$ increase would correspond to sea water temperature below the freezing point, which is impossible. Second, as showed by Chappell and Shackleton (1986), the Pacific record may be considered as a global sea water $\partial^{18}O$ record for that period. If the benthic foraminifera are in place in the Norwegian Sea sediment cores, their isotopic composition would correspond to deep water tempe- rature significantly warmer than today. The reconstruction of the deep water circulation in the North Atlantic (Duplessy et al.,1987) indicates the sinking of cold surface water at high latitudes, which fed the deep North Atlantic basin and contributed to the slow renewal of the deep water in the Norwegian Sea. Since temperature is a conservative tracer for deep waters, the presence of warm deep water in the Norwegian basin seems rather unlikely under these hydrological conditions. The third explanation, which we favor, is that the rare benthic foraminifera shells found in the sediment levels from stage 2 are not in place and that the Norwegian Sea was entirely devoid of benthic life at the peak of the glaciation between about 25,000 and 11,000 yr B.P., in agreement with previous observations of Streeter et al. (1982). The presence of such a barren zone for benthic foraminifera may result from the weak renewal and the poor oxygenation of the water in the deep Norwegian basin (Jansen et al.,1983; Sejrup et al., 1984) as a consequence of the stratification caused by the permanent ice cover and also from the concomitant low surface productivity which results in a reduction of the availability of food at the sediment surface (Belanger, 1982).

8. A Record of the Global Ocean $\partial^{18}O$ Variations: Major Consequences.

Our data demonstrate that the deep waters of the Norwegian Sea remained at quasi-constant low temperature over the entire last climatic cycle. This permanent cold temperature is due to the process of deep water formation which occurred in the Norwegian basin during isotope stages 1, 4 and 5 and to the proximity of the North Atlantic source of deep water during isotope stages 2 and 3.

The benthic calcite $\partial^{18}O$ record from the Norwegian Sea therefore reflects solely the $\partial^{18}O$ variations of the global ocean. Unfortunately this record is not complete, because benthic foraminifera are absent during isotope stage 2. Moreover, its accuracy is low during isotope stage 3, because the sedimentation was almost nil when sea ice was present so that the sedimentary record of one single core is incomplete. By contrast, the benthic $\partial^{18}O$ record of Pacific core V19-30 exhibits a good sequence reflecting only the ocean water $\partial^{18}O$ variations during the glaciation. We therefore obtained a complete record of the variations of the global ocean water $\partial^{18}O$ by piecing the isotopic record of core V19-30 for stages 2 and 3 with the stack benthic record of the Norwegian Sea for stages 1, 4 and 5. The resulting record is reported in Table 4 and in Fig. 10 as departure in per mil from the modern global ocean $\partial^{18}O$ as a function of time. It may be used as the reference record corresponding to $\partial^{18}O$ variations of a water mass which experienced no temperature variations during the last climatic cycle. By comparing it to the benthic record of other oceanic basins, it would provide a method to estimate the deep water temperature variations over the last climatic cycle (Labeyrie et al., 1987).

The record of global ocean $\partial^{18}O$ variations may also be used to estimate sea surface hydrological conditions. However the problem is

TABLE 4: Oxygen isotopic ratio of the mean ocean water
as a function of time during the last 135,000 years

Age kyr	mean sea water ∂w	Age kyr	mean sea water ∂w	Age kyr	mean sea water ∂w
0	0,00	46	0,59	91	0,20
1	-0,02	47	0,57	92	0,23
2	-0,01	48	0,58	93	0,29
3	0,01	49	0,44	94	0,28
4	0,01	50	0,33	95	0,17
5	0,05	51	0,38	96	0,10
6	0,10	52	0,45	97	0,10
7	0,17	53	0,53	98	0,08
8	0,27	54	0,56	99	0,02
9	0,32	55	0,57	100	0,09
10	0,37	56	0,58	101	0,18
11	0,42	57	0,57	102	0,22
12	0,48	58	0,43	103	0,24
13	0,48	59	0,20	104	0,34
14	0,88	60	0,35	105	0,48
15	0,91	61	0,45	106	0,56
16	1,05	62	0,48	107	0,51
17	1,01	63	0,69	108	0,43
18	0,94	64	0,70	109	0,41
19	0,95	65	0,71	110	0,47
20	1,01	66	0,67	111	0,50
21	1,00	67	0,63	112	0,46
22	0,95	68	0,55	113	0,45
23	0,80	69	0,56	114	0,42
24	0,70	70	0,49	115	0,41
25	0,51	71	0,45	116	0,30
26	0,56	72	0,40	117	0,22
27	0,59	73	0,40	118	0,11
28	0,60	74	0,44	119	0,05
29	0,65	75	0,47	120	-0,06
30	0,67	76	0,39	121	-0,15
31	0,65	77	0,26	122	-0,14
32	0,63	78	0,17	123	-0,11
33	0,68	79	0,08	124	-0,05
34	0,63	80	0,08	125	-0,03
35	0,55	81	0,15	126	0,00
36	0,48	82	0,25	127	0,09
37	0,55	83	0,18	128	0,20
38	0,61	84	0,09	129	0,35
39	0,67	85	0,15	130	0,49
40	0,67	86	0,29	131	0,61
41	0,64	87	0,40	132	0,73
42	0,58	88	0,43	133	0,85
43	0,63	89	0,37	134	0,98
44	0,64	90	0,28	135	1,10
45	0,61				

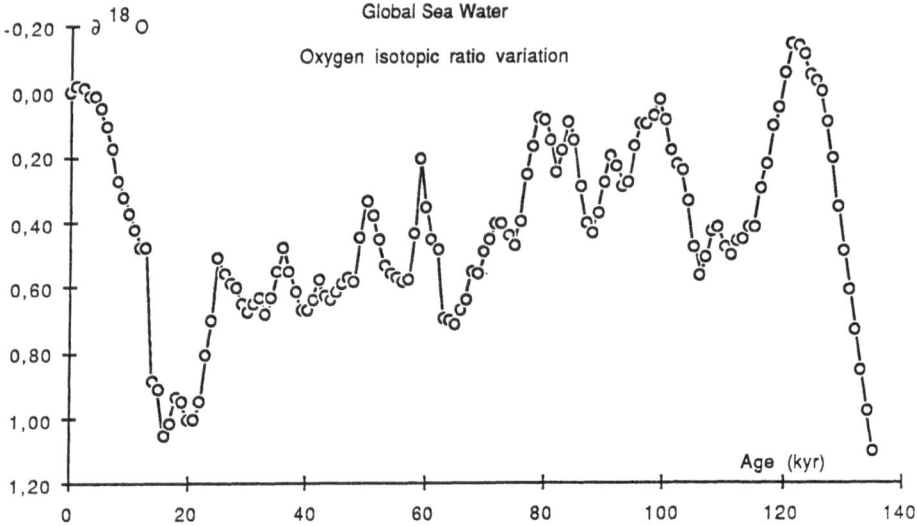

Figure 10 : Variations of the oxygen isotopic composition of the global ocean water during the last climatic cycle obtained by piecing the stacked benthic Norwegian Sea record of isotope stages 1, 4 and 5 with the benthic record of Pacific core V 19-30 for glacial isotope stages 2 and 3.

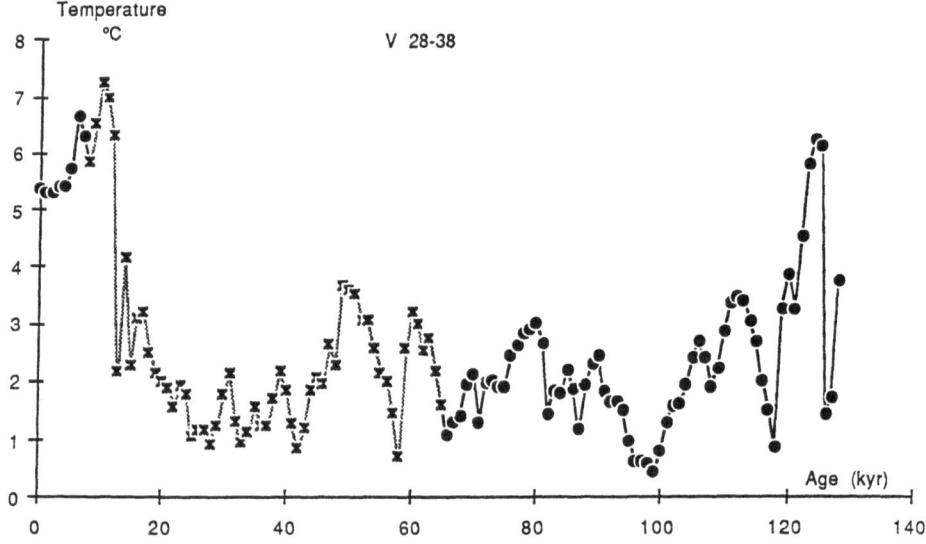

Figure 11 : Estimate of sea surface temperature variations in the Norwegian Sea calculated from the $\partial^{18}O$ difference between the planktonic record of core V 28-38 and the global sea water $\partial^{18}O$ record. Between 9 and 65 kyr, no constraints are available for surface salinity and $\partial^{18}O$ variations and sea surface temperature are probably overestimated as a consequence of melting of ice depleted in ^{18}O (this part of the record is drawn in dashed lines).

more complicated than the interpretation of the benthic record because the $\partial^{18}O$ planktonic record depends on three variables: global $\partial^{18}O$ variations, sea surface temperature (at the season and depth at which the foraminifera have lived), and local surface water $\partial^{18}O$ changes, which may be due to local changes either in the evaporation/precipitation ratio or in the rate of exchange between surface and deep waters. If micropaleontological transfer functions can be used to obtain an independent record of sea surface temperature (SST) variations, the comparison of the planktonic $\partial^{18}O$ record with the global ocean $\partial^{18}O$ record and with the SST record should provide a method to estimate the local sea surface $\partial^{18}O$ values and therefore the sea surface salinity.

In the Norwegian Sea, where micropaleontological transfer functions are not accurate, estimates of SST could be obtained when some constraints exist on surface salinity. As an example, we compared the $\partial^{18}O$ record of *N. pachyderma* in core V 28-38 with the global sea water $\partial^{18}O$ record. In order to compare these records, we estimated by linear interpolation the $\partial^{18}O$ values of *N. pachyderma* each 1,000 years. We assumed that the $\partial^{18}O$ difference between surface and deep water has remained negligible during the whole time the Norwegian Sea was an active source of deep water. The variations of sea surface $\partial^{18}O$ value may thus be estimated by adding the modern value to the global sea water $\partial^{18}O$ record. The sea surface temperature has then been calculated by using the paleotemperature equation (Shackleton, 1974). Results are reported in Fig.11. The estimate for modern conditions (5.5°C) falls right within the range of presently measured SST (3°C during winter and 8°C during summer). The estimated temperatures exhibit values similar to the present ones only during the peak of isotopic substage 5e, and a major temperature drop occurred by the end of substage 5e. During the remainder of stage 5, the temperatures were in the range 1-3°C, as expected from the presence of coccoliths in the sediment from this time. During the glaciation, we have no constraint on the surface salinity and $\partial^{18}O$ values and our model is not valid any more. The estimated temperatures are probably too high, since salinity and $\partial^{18}O$ are expected to decrease, as a consequence of the presence of a quasi-permanent ice cover over the Norvegian Sea. High temperature estimates are obtained for the Holocene, when our model is valid, with temperature higher than today about 7,000 years ago, in agreement with the observations of Kellogg (1976).

9.   Conclusions

Oxygen and carbon isotope analyses of planktonic and benthic foraminifera from five Norwegian Sea sediment cores demonstrate that the Norwegian Sea was an active area of deep water formation not only during full interglacial conditions (isotopic stages 1 and 5e) but also during the early part of the glaciation (isotope stages 4 and 5a-5d).

As sinking of surface water to the depth is linked to its cooling during winter, the Norwegian Sea deep water temperature was close to the freezing point during the time intervals 0-8 kyr B.P. and 65-128 kyr B.P.. Consequently the oxygen isotope record of this period is a record of the global sea water $\partial^{18}O$ variations due to the growth and decay of continental ice sheets.

A complete record of the global sea water $\partial^{18}O$ variations during the last climatic cycle has been established by piecing the stacked Norwegian Sea benthic record of stages 1, 4, and 5 with the benthic record of Pacific core V 19-30 for glacial stages 2 and 3.The resulting record may be used to extract either the temperature signal present in benthic records from other oceanic basins or the variations in surface hydrology recorded by the oxygen isotope variations of planktonic foraminifera.

C.F.R. Contribution N° 870.

REFERENCES

Bard, E., M. Arnold, J. Duprat, J. Moyes, and J.C. Duplessy, Recons-
    truction of the last deglaciation: Deconvolved records of ∂18O
    profiles, micropaleontological variations and accelerator mass
    spectrometric 14C dating, *Climate Dynamics*, 1, 101-112, 1987.

Belanger, P.E., Paleo-oceanography of the Norwegian Sea during the past
    130,000 years: coccolithophorid and foraminiferal data, *Boreas*, 11,
    29-36, 1982.

Belanger, P.E., and S.S. Streeter, Distribution and ecology of benthic
    foraminifera in the Norwegian-Greenland sea,*Mar. Micropal*, 5, 401-
    428, 1980.

Blanc, P.L. and J.C. Duplessy, The deep water circulation during the
    Neogene and the impact of the Messinian salinity crisis, *Deep Sea
    Res.*, 29, 1391-1414, 1983.

Boyle, E.A. and L.D. Keigwin, Deep circulation of the North Atlantic
    over the last 200,000 years: Geochemical evidence, *Science*, 218,
    784-787, 1982.

Chappell J. and N.J. Shackleton, Oxygen isotopes and sea level, *Nature*,
    324, 137-140, 1986

Dodge, E.D., R.G. Fairbanks, L.K. Benninger, and F. Maurasse,
    Pleistocene sea level from raised coral reefs of Haiti, *Science*,
    219, 1423-1425, 1983.

Duplessy, J.C., Isotope studies, in *Climatic Change*, edited by J.
    Gribbin, Cambridge University Press, Cambridge, 46-67, 1978.

Duplessy, J.C., Circulation des eaux profondes nord atlantiques au cours du dernier cycle climatique, *Bull. Inst. Géol. Bassin d'Aquitaine*, 31, 379-391, 1982.

Duplessy, J.C., C. Lalou, and A.C. Vinot, Differential isotopic fractionation in benthic foraminifera and paleotemperature reassessed, *Science*, 168, 250-251, 1970.

Duplessy , J.C., L. Chenouard, and F. Vila, Weyl's theory of glaciation supported by isotopic study of Norwegian core K-11, *Science*, 188, 1208-1209, 1975.

Duplessy, J.C., J. Moyes and C. Pujol, Deep water formation in the North Atlantic Ocean during the last ice age, *Nature*, 286, 479-482, 1980.

Duplessy, J.C., G. Delibrias, J.L. Turon, C. Pujol, and J. Duprat, Deglacial warming of the northeastern Atlantic ocean: correlation with the paleoclimatic evolution of the european continent, *Palaeogeogr., Palaeoclimat., Palaeoecol.*, 35, 121-144, 1980.

Duplessy, J.C. and N.J. Shackleton, Response of global deep-water circulation to the Earth's climatic change 135,000-107,000 years ago, *Nature*, 316, 500-507, 1985.

Duplessy, J.C., M. Arnold, P. Maurice, E. Bard, J. Duprat and J. Moyes, Direct dating of oxygen-isotope record of the last deglaciation by C-14 accelerator mass spectrometry, *Nature*, 320, 350-352, 1986.

Duplessy, J.C., N.J. Shackleton, R.K. Matthews, W. Prell, W.F. Ruddiman, M. Caralp, and C.H. Hendy, 13C record of benthic foraminifera in the last interglacial ocean: Implications for the carbon cycle and the global deep water circulation, *Quat. Res.*, 21, 225-243, 1984.

Duplessy, J.C., N. J. Shackleton, R. Fairbanks, L. Labeyrie, D. Oppo and N. Kallel, Deep water source variations during the last climatic cycle and their impact of the global deep water circulation, *Paleoceanography* (submitted), 1987.

Emiliani, C., Pleistocene Temperatures, *J. Geol.*, 63, 538-578, 1955.

Holtedahl, H., Geology and paleontology of Norwegian sea bottom cores, *J. Sed. Petrol.*, 29, 16-29, 1959.

Imbrie, J., J.D. Hays, D.G. Martinson, A. Mc Intyre, A. Mix, J. J. Morley, N. Pisias, W. Prell, and N.J. Shackleton, The orbital theory of Pleistocene climate:Support from a revised chronology of the late marine $\partial18O$ record, in *Milankovitch and Climate* edited by A. Berger et al., D. Reidel, Hingam, Mass., 1984

Jansen, E., H.P. Sejrup, T. Fjaeran, M. Hald, H. Holtedahl and O. Skarbo, Late Weichselian paleoceanography of the southeastern Norwegian Sea, *Norsk Geologisk Tidsskrift*, 63, 117-146, 1983

Kellogg, T.B., Late Quaternary climatic changes: evidence from cores from Norwegian and Greenland Seas, *Geol. Soc. Am. Memoir*, 145, 77-110, 1976.

Kellogg, T.B., Paleoclimatology and paleo-oceanography of the Norwegian and Greenland seas: Glacial-interglacial contrasts, *Boreas*, 9, 115-137, 1980.

Kellogg, T.B., N.J. Shackleton, and J.C. Duplessy, Planktonic fora-
    miniferal and oxygen isotopic stratigraphy and paleoclimatology of
    Norwegian sea deep sea cores, *Boreas*, 7, 61-73, 1978.

Labeyrie, L.D. and J.C. Duplessy, Changes in the oceanic C13/C12 ratio
    during the last 140,000 years: high latitude surface water records,
    Palaeogeogr., Palaeoclimatol., Palaeoecol., 50, 217-240, 1985.

Labeyrie, L. D., J.C. Duplessy, and P.L. Blanc, Deep water formation and
    temperature variations over the last 125,000 years, *Nature*, 327,
    477-482, 1987.

Mackensen, A., H.P. Sejrup, and E. Jansen, The distribution of living
    benthic foraminifera on the continental slope and rise off southwest
    Norway, *Mar. Micropal.*, 9, 275-306, 1985.

Mix, A. and R.G. Fairbanks, North Atlantic surface-ocean control of
    Pleistocene deep ocean circulation, *Earth Planet. Sci. Lett.*, 73,
    231-243, 1985.

Paterne, M., F. Guichard, J. Labeyrie, P.Y. Gillot and J.C. Duplessy,
    Tyrrhenian sea tephrochronology of the oxygen-isotope record for the
    past 60,000 years, *Mar. Geol.*, 72, 259-285, 1986.

Reid, J.L. and R.L. Lynn, On the influence of the Norwegian-Greenland
    and Weddell seas upon the bottom waters of the Indian and Pacific
    oceans, *Deep Sea Res.*, 18, 1063-1088, 1971.

Sarnthein, M., H. Erlenkeuser, R. von Grafenstein, and C. Schroeder,
    Stable-isotope stratigraphy for the last 750,000 years: "Meteor"
    core 13519 from the eastern equatorial Atlantic, Meteor
    Forschungergebnisse, Reihe C , 38, 9-24, 1984.

Savitzky, A. and J. E. Golay, Smoothing and differenciation of data by
    simplified least squares procedure, *Anal.Chem.*, 36, 1627-1639, 1964.

Sejrup, H.P., E. Jansen, H. Herlenkeuser, and H. Holtedahl, New faunal
    and isotopic evidence of the late Weichselian-Holocene oceanographic
    changes in the Norwegian Sea, *Quat. Res.*, 21, 74-84,1984.

Shackleton, N.J., Attainment of isotopic equilibrium between ocean water
    and the benthonic foraminifer genus Uvigerina : isotopic changes in
    the ocean during the last glacial, *Colloque International du Centre
    National de la Recherche Scientifique,* N° 219, 203-210, 1974.

Shackleton, N.J., Carbon-13 in Uvigerina: Tropical rainforest history
    and the equatorial Pacific carbonate dissolution cycle, in *The Fate
    of fossil fuel CO2 in the Oceans*, edited by N. R. Anderson and A.
    Malahoff, pp 401-428, Plenum, New York, 1977.

Shackleton, N.J., and M.B. Cita, Oxygen and carbon isotope stratigraphy
    of benthic foraminifera at site 397: Detailed history of climatic
    change during the late Neogene, *Initial Reports of the Deep Sea
    Drilling Project*, 47, 433-445, 1979.

Shackleton, N.J., J. Imbrie, and M. Hall, Oxygen and Carbon isotope
    record of east Pacific core V 19-30: Implications for the formation
    of deep water in the Late Pleistocene North Atlantic, *Earth and
    Planet. Sci. Lett.*, 65, 233-244, 1983.

Shackleton, N.J. and N. D. Opdyke, Oxygen isotope and paleomagnetic stratigraphy of equatorial Pacific core V28-238: Oxygen isotope temperatures and ice-volumes on a 105 year and a 106 year scale, *Quat. Res.*, 3, 39-55, 1973.

Streeter, S.S. , P.E. Belanger, T.B. Kellogg, and J.C. Duplessy, Late Pleistocene paleo-oceanography of the Norwegian-Greenland Sea: Benthic foraminiferal evidence, *Quat. Res.*, 18, 72-90, 1982.

Zahn, R., K. Winn, and M. Sarnthein, Benthic foraminiferal $\partial 13C$ and accumulation rates of organic carbon: Uvigerina peregrina group and Cibicidoides wuellerstorfi, *Paleoceanography*, 1, 27-42, 1986.

Table 1: Core K-11

depth (cm)	N. pachyderma ∂18	∂13	Cibicides wuellerstorfi ∂18	∂13	Oridorsalis tenei ∂18	∂13	Benthic ∂18
0	3,02	0,49	3,83	1,30	4,31	-0,61	4,57
3			3,89	1,36			4,53
5	3,06	0,33					
8			4,11	1,21	4,48	-0,84	4,80
10	3,22	0,42					
12			4,01	1,25	4,76	-1,07	4,89
15	3,60	0,51					
16	3,74	0,43					
17			3,92	1,13			4,56
18	3,25	0,55					
20	3,19	0,48	3,84	1,17	4,66	-1,59	4,75
23	3,77	0,33					
25	3,55	0,09					
26			3,85	1,19	4,76	-1,59	5,12
29			3,90	1,27	4,96	-1,50	5,32
30	4,59	0,14					
32			3,99	1,32	4,77	-1,56	5,13
35	4,71	0,14	3,92	1,23	4,71	-1,49	5,07
38			3,93	1,26	4,80	-1,79	5,16
40	4,57	0,11	3,91	1,17			
45	4,73	0,22	3,91	1,22	5,00	-2,20	5,36
50	4,65	0,10	3,94	1,28	5,01	-2,04	5,37
53			3,90	1,32			
56	4,75	0,08	3,91	1,17	4,81	-1,49	5,17
59	4,53	0,31	3,94	1,21	4,59	-1,89	4,95
60	4,58	0,31					
65	4,37	0,21	4,25	1,21	4,68	-1,73	5,04
68	4,52	0,28	3,98	1,27	5,06	-1,60	5,42
70	4,46	0,33	3,88	1,18	4,71	-1,73	5,07
73	4,53	0,39	3,95	1,21	4,95	-1,49	5,31
75	4,52	0,22					
80	4,58	0,47	4,12	1,44	5,01	-1,54	5,37
83			3,91	1,30			
85	4,16	0,27					
90	4,35	0,30	3,92	1,21	4,83	-1,53	5,19
94			3,91	1,19			
95	4,57	0,22					
100	4,48	0,40	3,88	1,21	4,96	-1,34	5,32
103			3,89	1,28			
105	4,46	0,41					
106	4,68	0,39	3,85	1,26	4,94	-1,15	5,30
110	4,09	0,36					
113			3,94	1,74	4,71	-1,31	5,07
115	3,74	0,25					
120	4,56	0,37					
121			4,00	1,42	4,88	-1,37	5,24
124			3,93	1,26			
125	4,36	0,37					
126			3,82	1,26			
130	4,65	0,46	4,07	1,30	5,08	-1,10	5,44

Table 1: Core K-11 (cont.)

depth (cm)	N. pachyderma ∂18	N. pachyderma ∂13	Cibicides wuellerstorfi ∂18	Cibicides wuellerstorfi ∂13	Oridorsalis tenei ∂18	Oridorsalis tenei ∂13	Benthic ∂18
133			4,08	1,27	4,85	-1,20	5,21
135	4,27	0,47					
138			4,04	1,46	4,74	-1,30	5,10
140	4,39	0,39					
144			3,98	1,19	4,91	-1,10	5,27
145	4,65	0,55					
146			4,10	1,30			
148			4,00	1,22			
149					5,03	-1,12	5,39
150	4,06	0,17					
152			3,94	1,28			
154			3,97	1,27	4,63	-1,43	4,99
155	4,48	0,44					
157			3,84	1,22			
160	4,28	0,40	4,04	1,30	4,65	-1,33	5,01
165	4,22	0,23	3,87	1,26	4,86	-1,76	5,22
170	4,42	0,30					
171			4,02	1,26	4,75	-1,42	5,11
173			4,02	1,27			
175	4,23	0,28					
176			4,00	1,39	4,80	-1,42	5,16
179			3,95	1,33	4,79	-1,50	5,15
180	4,34	0,09					
182			3,99	1,30	4,86	-1,35	5,22
185	4,25	-0,04	4,10	1,37	4,74	-1,45	5,10
187	4,29	0,10	3,92	1,21	4,74	-1,45	5,10
190	4,46	0,20	3,84	1,24	4,89	-1,66	5,25
194	4,66	0,13	3,91	1,27	4,46	-1,33	4,82
195	4,48	0,02					
196			4,03	1,28	4,90	-1,78	5,26
198			4,22	1,50	4,60	-1,79	4,96
200	4,63	0,27					
201			4,50		4,86	-1,25	5,18
202			4,52	1,36			5,16
204			4,65	1,37			5,29
208			4,41	1,32			5,05
210	4,55	0,65	4,42	1,34			5,06
214			4,31	1,44			4,95
215	4,32	0,56					
218			4,24	1,52			4,88
220	4,19	0,59					
224			4,29	1,36			4,93
228			4,24	1,25			4,88
230	3,99	0,51					
234			4,19	1,20			4,83
238			4,17	1,62			4,81
240	3,98	0,68	4,11	1,63			4,75
242			3,92	1,40			4,56
244			3,98	1,61			4,62
245			4,26	1,22			4,90

Table 1: Core K-11 (cont.)

depth (cm)	N. pachyderma ∂18	∂13	Cibicides wuellerstorfi ∂18	∂13	Oridorsalis tener ∂18	∂13	Benthic ∂18
248			3,86	1,54			4,50
250	4,25	0,62					
252			4,29	1,33			4,93
255			4,36	1,31			5,00
258			4,25	1,40			4,89
260	4,18	0,54					
262			4,08	1,46			4,72
264			4,11	1,35			4,75
265			4,09	1,03			4,73
268			4,05	1,41	4,26	-0,76	4,69
270	4,29	0,72					
272			4,05	1,33			4,69
274			3,86	1,18	4,35	-0,80	4,50
275	4,07	0,50					
279			3,70	1,08			4,34
280	3,42	0,45					
282	3,46	0,37	3,85	1,19			4,49
284			3,76	1,13	4,11	-1,21	4,44
285	3,10	0,10	3,79	1,10			4,43
288			3,82	1,16			4,46
290	3,83	0,26					
292	3,85	0,23	3,82	1,17			4,46
298	3,56	0,15	3,70	1,02			4,34
300	3,35	0,06					
303					5,24	-2,33	5,60
305	4,75	0,08					
310	4,90	0,06					

Table 1: Core V 27-60

depth (cm)	N. pachyderma left ∂18	∂13	Cibicides wuellerstorfi ∂18	∂13	Oridorsalis tener ∂18	∂13	Benthic ∂18
12	2,84	0,40	3,88	1,24	4,15	-0,83	4,52
14	2,71		3,65	1,15	4,05	-0,77	4,35
20	2,82	0,34	3,90	1,28	4,26	-0,72	4,53
30	2,87	0,57	3,97	1,24	4,24	-0,70	4,61
34,5	2,81	0,67	3,81	1,31	4,15	-0,63	4,48
40	2,76	0,46	3,99	1,34	4,11	-0,93	4,55
44	2,67	0,62	3,67	1,31	4,06	-0,84	4,37
52	2,81	0,44	3,91	1,19			4,55
55	2,7	0,45	4,12	1,20	4,21	-0,87	4,67
60	2,69	0,35	3,85	1,13			4,49
70	2,5	0,22	3,88	1,11	4,28	-1,11	4,58
74	2,42		4,12	1,12	4,24	-0,71	4,68
78	2,65	0,26	4,17	1,10			4,81

Table 1: Core V 27-60 (Cont.)

depth (cm)	N. pachyderma left		Cibicides wuellerstorfi		Oridorsalis tener		Benthic
	$\partial 18$	$\partial 13$	$\partial 18$	$\partial 13$	$\partial 18$	$\partial 13$	$\partial 18$
80	2,8	0,06	4,02	1,05	4,38	-1,08	4,70
85	2,63	0,09	4,11	0,97			4,75
90	2,85	0,18	4,12	0,91	4,35	-1,32	4,74
99	2,95	0,17	4,20	0,97	4,46	-1,28	4,82
104	2,82	0,20	4,24	1,25			4,88
110					4,62	-1,06	4,98
117	2,93	0,33			4,55	-0,84	4,92
120	3,56	0,36					
127					4,42	-0,83	4,81
137,5	3,65	0,54	4,17	0,95			4,81
140	4,07	0,34					
160	3,62	-0,06					
170	4,34	-0,02					
186					4,71	-1,40	5,07
190	4,84	0,05	4,05	1,10	4,85	-1,61	5,21
200	4,72	0,02	4,09	1,20			
210	4,62	0,04					
220	4,59	0,00					
230	4,56	-0,11					
240	4,43	-0,05					
249	4,57	0,09					
259	4,74	0,22			4,91	-1,70	5,27
270	4,53	0,13					
280	4,39	0,08					
291	4,43	0,12					
302	4,59	0,38					
310	4,5	0,31					
320	4,66	0,48					
332	4,47	0,23			4,76	-1,34	5,12
340	4,31	0,01			4,50	-1,42	4,86
350	4,13	0,07			4,92	-1,58	5,28
360	3,88	0,18					
370	3,98	-0,09			4,82	-1,99	5,18
375	4,45	-0,04					
380	4,38	0,26					
385	3,2	-0,17					
390	3,94	-0,14			4,84	-2,16	5,20
395	4,67	0,03					
401	4,61	0,33	4,41	1,17			5,05
403			4,64				5,28
405	4,38	0,23	4,55	1,22			5,19
408			4,42	1,29			5,06
410	4,49	0,46	4,47	1,26			5,11
415	4,28	0,45	4,50	1,29			5,14
417			4,48	1,32			5,12
420	4,33	0,64	4,22	1,47			4,86
425	4,16	0,60	4,28	1,52			4,92
428			4,49	1,77			5,13

Table 1: Core V 27-60 (Cont.)

depth (cm)	N. pachyderma left $\partial18$	$\partial13$	Cibicides wuellerstorfi $\partial18$	$\partial13$	Oridorsalis tener $\partial18$	$\partial13$	Benthic $\partial18$
431	3,96	0,41	4,30	1,43			4,94
435	3,83	0,30	3,93	1,32			4,57
437			4,02	1,27			4,66
440	3,49	0,53	3,73	1,22			4,37
445			4,25	1,57			4,89
446	4,15	0,65	3,72	1,33			4,36
450	4,19	0,60	4,19	1,46			4,83
458			3,86	1,26			4,50
461	4,17	0,61	3,80	1,26			4,44
465	4,26	0,64	4,16	1,33			4,80
467			4,16	1,29			4,80
475	3,96	0,44	4,35	1,31			4,99
480	3,94	0,37	4,40	1,24			5,04
485	3,82	0,34	4,37	1,16			5,01
487,5			4,29	0,99			4,93
492	3,81	0,33	4,34	1,23			4,98
495	4,28	0,69	4,34	1,30			4,98
498			4,27	1,10			4,91
500	4,34	0,66	4,35	1,19			4,99
505	4,38	0,74	3,98	1,23			4,62
508			4,12	1,08	4,29	-0,51	4,71
510	3,73	0,52	4,15	1,17			4,79
515	2,81	0,20	3,80	1,10			4,44
520			3,81	0,99			4,45
525	2,84	0,17	3,94	1,07			4,58
527,5			3,70	0,89			4,34
532	3,11	0,29	3,80				4,44
532			3,73	0,87	3,85	-0,75	4,29
535	2,8	0,07	3,68	0,82			4,32
536	2,56	0,01	3,46	0,87	3,60		4,03
545	2,72	0,03	3,88	0,94			4,52
548			3,75	0,88			4,39
552	2,75	-0,07	3,76	0,89			4,40
555	2,85						
560	3,28	0,13	3,91	0,69			4,55
565	4,3	0,28					
570			4,10	1,19			4,74
575	3,61	0,18					
580	4,27	-0,14					
584	4,17	-0,11					
591	4,13	-0,13					
610	4,23	-0,04					
620	3,99	-0,15					
640	4,78	0,11			5,29	-1,70	5,65
650	4,7	0,14			5,17	-1,86	5,53
660	4,56	0,04			5,07	-1,86	5,43
670	4,47	0,10			5,17	-1,81	5,53
680	4,6	0,02			5,13	-1,92	5,49

Table 1: Core V 27-86

Depth (cm)	N. pachyderma left ∂18	∂13	Cibicides wullerstorfi ∂18	∂13	Oridorsalis tener ∂18	∂13	Benthic ∂18
1	2,68	0,27	3,95	1,39	4,25	-0,86	4,60
5	2,67	0,18	4,07	1,36	4,21	-0,76	4,64
9	2,54	0,34			4,30	-0,83	4,66
15	2,63	0,35	4,08	1,31	4,27	-0,87	4,68
19	3,04	0,26	3,94	1,21	4,43	-0,97	4,69
30	4,30						
40	4,47	0,19			4,92	-2,14	5,28
50	4,72	0,08					
55					5,34	-1,98	5,70
60	4,71	0,04					
70	4,74	-0,07			4,77	-1,94	5,13
80	4,55	0,11					
100	4,68	0,03			4,65	-1,90	5,01
110	4,36	0,22					
120	4,25	0,12					
130	4,52	0,28			4,74	-1,36	5,10
140	4,38	0,42			4,85	-1,18	5,21
150	4,10				4,76	-1,60	5,12
160	4,38	-0,04			4,60	-1,59	4,96
169	3,40	0,38	3,89	1,28			4,53
210	3,64						
215	4,52	-0,10	4,36	1,33			5,00
220	4,34	0,15	4,33	1,52	4,53	-1,21	4,93
225	4,46	0,07	4,69	1,45			5,33
230	4,63	-0,01	4,35	1,27	4,72	-1,24	5,04
235	4,51	0,27	4,49	1,24	4,70	-1,02	5,10
240	4,30	0,40	4,04	1,40			4,68
246	3,78		4,48	1,54			5,12
250	3,92	0,39	4,26	1,36	4,61	-1,07	4,94
255	4,05	0,51					
260	4,39	0,54	4,27	1,57	4,72		4,91
265	4,16	0,68					
270	4,02	0,47	3,87	1,43	4,07	-1,13	4,47
275	4,20	0,53					
280	3,78	0,21	4,12	1,24			4,76
285	3,84	0,32	4,15	1,27			4,79
289	3,75	0,23	4,41	1,20			5,05
295	4,67	0,51	4,60	1,29			5,24
300	4,45	0,62	3,98	1,38			4,62
305	3,87	0,38	4,03	1,31			4,67
310	3,28	0,28	4,23	1,27			4,87
315	3,08	0,16	3,78	1,29			4,42
321	3,26	0,18	3,84	1,17			4,48
325	3,02	-0,08	3,81	1,15			4,45
330	4,04	0,19					
335	4,13	0,23	3,98	1,21			4,62
340	4,14	0,28	3,84	1,21			4,48
360	3,92	0,44					

Table 1: Core V 28-38

depth (cm)	N. pachyderma left ∂18	∂13	Cibicides wuellerstorfi ∂18	∂13	Oridorsalis tener ∂18	∂13	Benthic ∂18
48	2,94	0,59	3,90	1,30	4,16	-0,54	4,53
58	2,90	0,50	3,89	1,27	4,36	-0,63	4,62
69	2,72	0,40	3,94	1,32	4,32	-0,92	4,63
80	2,90	0,55	3,97	1,22	4,35	-0,93	4,66
92	3,19	0,43			4,39	-0,69	4,75
101	2,84	0,40					
112	2,84	0,40			4,45	-0,96	4,81
119	2,98	0,27			4,36	-0,95	4,72
132	3,24	0,51			4,66	-1,00	5,02
142	4,34	0,45			4,57	-1,13	4,93
148					4,71	-1,34	5,07
152	4,11	0,25			4,54	-1,88	4,90
158	4,33	0,08			4,47	-2,07	4,83
164	4,63	0,14			4,60	-1,20	4,96
172	4,50	0,17					
175					4,57	-1,03	4,93
184	4,79	0,12			4,63	-1,21	4,99
192	4,84	0,17					
194					4,54	-1,20	4,90
197	4,55	0,34					
204	4,53	0,11			4,29	-1,60	4,65
208	4,55	0,20			4,58	-0,93	4,94
212	4,68	0,31					
215					4,79	-1,49	5,15
218	4,40	0,34					
222	4,72	0,43			4,39	-1,08	4,75
228	4,73	0,47			4,68	-1,14	5,04
232	4,39	0,49			4,44	-0,93	4,80
236	4,52	0,47			4,24	-1,29	4,60
248	4,38	0,52			4,50	-1,10	4,86
253	4,68	0,44			4,93	-1,23	5,29
258	4,56	0,50			4,58	-1,16	4,94
262	4,65	0,45					
268	4,65	0,50			4,70	-1,36	5,06
272	4,62	0,38			4,62	-1,28	4,98
278	4,63	0,39			4,60	-1,31	4,96
282	4,54	0,31			4,45	-1,21	4,81
294	4,42	0,29			4,57	-1,41	4,93
302	4,38	0,18			4,41	-1,52	4,77
308	4,38	0,17					
314	4,08	0,09			4,82	-1,53	5,18
318	4,32	0,17			4,80	-1,62	5,16
323	4,06	0,08			4,64	-1,49	5,00
328	3,83	0,25			4,48	-1,23	4,84
336	3,67	0,05			4,69	-1,54	5,05
342	3,96	0,27			4,52	-1,43	4,88
352	4,06	0,38			4,63	-1,25	4,99
358	4,19	0,37			4,70	-1,37	5,06

Table 1: Core V 28-38 (Cont.)

depth (cm)	N. pachyderma left ∂18	∂13	Cibicides wuellerstorfi ∂18	∂13	Oridorsalis tener ∂18	∂13	Benthic ∂18
362	4,23	0,16					
368	4,33	0,30			4,68	-1,10	5,04
374	4,51	0,16			4,85	-1,41	5,21
378	4,10	0,14			4,83	-1,21	5,19
384	4,54	0,19			4,73	-1,07	5,09
391	4,39	0,12			4,18	-1,63	4,54
396			4,01	1,20			4,65
398	4,58	0,28			4,64	-1,65	5,00
402	4,00	-0,07			4,87	-1,14	5,23
408	3,73	-0,02	4,05	1,38	4,79	-1,44	4,92
414	4,59	0,21	4,34	1,26	4,93	-1,72	5,14
418	4,68	0,31			4,89	-1,31	5,25
425	4,48	0,75	4,46	1,44	4,94	-1,01	5,21
432	4,24	0,51	4,26	1,50	4,65	-1,24	4,96
434	4,24	0,74	4,28	1,68	4,57	-0,94	4,93
436	4,40	0,47	4,06	1,62	4,35	-0,99	4,71
438	4,13	0,69	4,11	1,63	4,51	-0,95	4,81
442	4,30	0,72	4,25	1,51	4,55	-1,04	4,90
445	3,91	0,37			4,61	-1,00	4,97
447			3,88	1,05			4,52
448	3,64	0,51	3,45	1,24	4,54	-1,10	
453	3,60	0,64	3,88	1,45			4,52
457			4,16	1,53			4,80
458	4,20	0,79			4,52	-1,00	4,88
462	4,02	0,35	4,17	1,35	4,43	-1,07	4,80
468	3,92	0,71					4,78
475	3,85	0,85	4,17	1,66	4,38	-1,17	4,83
478	4,33	0,77	4,28	1,74	4,37	-1,06	4,87
481	4,38	0,70	4,29	1,58	4,45	-0,99	4,76
486	3,94	0,59	4,12	1,56	4,40	-0,85	4,81
491	4,15	0,66	4,15	1,57	4,47	-1,01	4,69
498	4,23	0,79	3,73	1,52	4,33	-0,90	4,53
502	4,19	0,66	3,87	1,46	4,19	-1,17	4,68
508	4,10	0,78	3,91	1,47	4,45	-1,18	5,01
517	4,19	0,66			4,65	-1,17	4,87
518	4,29	0,79	4,18	1,44	4,55	-1,17	4,88
520					4,52	-1,01	4,91
521	4,14	0,59	4,27	1,54			4,66
522	4,12	0,82	4,02	1,36			4,72
527	3,88	0,46	4,05	1,23	4,39	-1,41	4,66
528	3,87	0,49	3,89	1,27	4,43	-1,12	4,46
532	3,88	0,50	3,81	1,14	4,11	-1,06	4,77
536	4,05	0,56	4,13	1,18			4,58
541	4,17	0,75			4,22	-0,81	4,72
543	3,62	0,49	4,04	1,22	4,40	-0,87	4,46
548	3,26	0,41	3,80	1,26	4,11	-0,87	4,36
552	3,33	0,42	3,74	1,19	3,98	-0,90	4,45
558	2,73	0,19	3,83	1,04	4,07	-0,82	4,47
562	2,70	0,10	3,78	1,00	4,15	-0,76	4,47
566	2,57	-0,01	3,77	0,96	4,17	-0,94	4,71
571	4,13	0,34			4,35	-1,26	4,73
573	4,05	0,39			4,37	-1,29	4,79
578	3,55	-0,07			4,43	-1,20	

Table 1: Core CH 77-07

depth	N. pachyderma left		Cibicides wuellerstorfi		Oridorsalis tener		Benthics
cm	∂18	∂13	∂18	∂13	∂18	∂13	∂18
0	3,78	0,51	3,86	1,10	4,20	-0,97	4,53
8			3,82	0,92	4,09	-0,95	4,46
10	3,74	0,54					
18			3,95	0,79	4,22	-0,93	4,59
20	3,81	0,76					
28			3,84	0,95	4,15	-1,00	4,50
30	3,72	0,53					
38			4,00	0,91			4,64
40	3,77	0,29					
48			4,00	0,81			4,64
50	4,10	0,29					
58			4,05	0,82	4,27	-1,96	4,66
60	3,95	0,23					
68			4,11	0,83	4,23	-1,26	4,67
70	3,92	-0,08					
78			4,10	0,77			4,74
80	3,92	0,08					
88			4,02	0,58	4,25	-1,45	4,64
90	4,24	0,04					
100	4,42	0,03					
105					4,66	-1,37	5,02
108					4,65	-1,44	5,01
110	4,27	-0,03					
118					4,94	-1,89	5,30
120	4,22	-0,24					
128	4,63	-0,10					
130	4,82	-0,02					
133	3,94	-0,03					
135	4,39	-0,09					
138	4,12	0,10					
140	4,04	-0,01					
143	4,52	0,04					
145	4,70	0,06					
150	4,74	-0,02					
160	4,61	0,12					
170	4,50	0,18					
180	4,56	0,23					
190	4,52	0,24					
200	4,74	0,38					
210	4,51	0,11					
220	4,39	0,05					
230	4,19	0,02					
240	4,55	-0,04					
250	4,32	0,03					
260	4,57	-0,01					
270	4,55	-0,04					
280	4,44	-0,05					
290	4,51	-0,02					
310	4,69	0,05					
330	4,46	0,07					
340	4,29	-0,04					
360	4,50	-0,02					
380	4,18	0,20					
390	4,35	0,17					

NUMERICAL MODELS OF CLIMATE

Hartmut Grassl
Forschungszentrum Geesthacht
Max-Planck-Straße 1
D-2054 Geesthacht

1. INTRODUCTION

The climate on earth is the result of many interactions between atmosphere, ocean, cryosphere, biosphere and lithosphere as well as interactions within a single compartment. At a given insolation the strong observed natural variability of climate parameters merely results from the drastically different response time of the compartments to changed boundary conditions.

Stimulated by extreme weather and climate conditions man started to establish a meteorological network in the 18th century leading to nowadays famous time series especially for temperature. The observations were meant as giving the climate statistics of an area after a sufficiently long time (some decades) once and for all. However, paleoclimatic evidence and these direct observations have led to a new climate definition: climate is the average weather after averaging over more than the expected ultimate weather forecast interval of a few weeks including all statistical parameters, thus also the probability for a distinct deviation from the mean. Since climate varies on all time scales also the time span used for averaging has to be specified. Climate defined in such a way accounts for the dominance of dynamics.

Being aware of the strong dynamics of climate it is clear that we need a continuous global observation network. At the same time we would - given such a global observing system - only be able to foresee any change in climate undisturbed by man, if strong periodicities accounting for a high percentage of variability are encountered. Since neither a sufficiently accurate fine mesh global observing system exists nor a known strong periodicity besides yearly and daily cycle could be derived for time scales below a few hundred years another tool for getting more insight has not only been created but also recently became a main research direction for climatologists: numerical models of global climate, although the ability to forecast climate under the conditions mentioned was not at all foreseeable. Another 'pressure' for modellers has to be realized: The present inability to separate in observations natural climate variability from man made climate change. Two main avenues have been followed in climate modelling:

- simple, mainly one-dimensional models for the understanding of distinct processes and basic interactions
- use of three-dimensional general atmospheric circulation models, developed by meteorological services, also for climate studies.

After shortly introducing the basic applications of modelling and the necessary equations in chapter 2, the main errors and drawbacks will be named in chapter 3, while the chapters 4 and 5 are devoted to results of numerical modelling both for the natural and disturbed climate system.

2. THE BASIS OF CLIMATE MODELLING

The basic laws necessary and the climate system compartments involved for distinct purposes in meteorology and climatology strongly depend on the time scale envisaged. This dependence will be shown firstly for meteorological and secondly for climate applications at growing time scales.

2.1 Meteorological Applications

1) For a short term (12 or 24 hours) forecast, of the pressure field neither irreversible processes in the atmosphere like radiation flux divergence, heat of condensation and turbulent heat fluxes play a dominant role nor changes at the boundaries ocean and land surface have to be introduced.

   Thus, the equation of motion and conservation of total mass often combined in a balance equation of potential vorticity are the only basic laws involved. No distinct vertical resolution is necessary.

2) For a 5-day weather forecast the irreversible processes become an important part as well as land surface parameter changes. The ocean surface temperature, however, may still be held constant at its typical seasonal values - or better - at its actual values at the start of the forecast.

   Now the basic thermodynamic laws, the radiative transfer equation and the equation of state (for the atmosphere simply the ideal gas law) have to be included, leading - besides the prognostic equation of motion - also to a prognostic equation for temperature. If the water cycle is also described in detail a prognostic water mass balance equation has to be solved.

3) For a 2 week weather forecast - still not possible at present, but on this side of the ultimate threshold for deterministic weather forecasting - the short term changes in the oceanic deck layer as well as surface water balance relations have to be added, thus not only bringing in the upper ocean layers but also more strongly the intrinsic interactions between soils and vegetation.

While it was rather simple to add a known basic equation under 1) and 2) the soil-vegetation-atmosphere interaction opens a new kind of difficulty: often poorly understood biological processes have to be formulated in mathematical equations.

2.2 Climate Applications

All the above meteorological applications were not directly answering questions for the main purpose of this contribution: climate models. Nevertheless 3) has pointed to the basic equations involved in both long term weather forecast and climate models. By enlarging the time interval we reach climate models. On one hand we thereby enhance difficulties because of the need to describe at least the global atmosphere and upper parts of the global ocean, on the other hand we no longer need an initial data set at the start of the computation very near to reality. The reason for the latter being both the failure of deterministic weather forecasting after a few weeks (because of the large impact of minute errors in the initial data on the final result due to non linear interactions) and the wish for climate statistics for the time interval desired.

The two main streams of climate research with respect to modelling (see WMO, 1984) presently are:

1) Understanding of year-to-year variations by using atmospheric general circulation models coupled to an upper ocean model. The ocean model has to reach the deepest permanent thermocline.

2) Decadal changes of climate caused by man's activities.

This modelling task enlarges both the climate system compartments - the entire ocean, deeper layers of soil, big parts of the cryosphere - and the number of mass balance equations for distinct substances, the latter describing the atmospheric chemistry and thus the trace substance composition of the atmosphere.

While task 1 is clearly tied to a better understanding of the natural climate system, task 2 is trying to isolate man's influence before the natural system on these time-scales is fully understood. Task 2 has also to rely on the scenarios of atmospheric composition change, thereby often loading climate model results with uncertainties from another part of 'forecasting', which is totally depending on extrapolation of man's energy use and thus also on political structures.

The base of climate models as compared to that of models of socio-economic development is rather clear:

- most equations are known
- the knowledge of the climate system is advanced enough to tell us the compartments to be handled for a distinct climate time interval
- empirical relations between vegetation cover and reflectivity of the surface are reasonably well established
- most important chemical reactions are also known. Despite this rather optimistic view the main error sources of climate models have to be discussed before results from different groups are presented in chapters 4 and 5.

3. ERRORS INHERENT TO CLIMATE MODELS

3.1 Numerical Errors

The basic physical equations are non linear thus analytic solutions do not exist. In turn, the numerical solution means integration of discretized differential equations in space and time. Thus, numerical errors inherent in any discretization are always present, although more advanced numerical techniques recently have reduced the impact on model results. However, a further strong reduction of this error is not easy, since a finer grid or mesh which would help also means - due to numerical instabilities otherwise imminent - a smaller time step, soon surmounting the capacity of any big computer. Halving the grid size increases the computer time in a three-dimensional climate model by a factor of 16.

3.2 Parameterization Errors

A second even larger error is also tied to the restricted computer capacity: the parameterization error. Because the basic laws in continuum physics are a closed

system only if applied unaveraged to volume elements smaller than the smallest turbulence element, any volume averaging of these equations leads to correlation products
of turbulent quantities with the need for a closure hypothesis. Many closure hypotheses for small scall turbulence with space scales of a few hundred meters and time
scales of half an hour exist, the simpler ones also used in climate models. However,
the typical grid size of a few hundred kilometers makes the parameterization problem
worse: all unresolved processes – not only small scall turbulence – have to be described in terms of mean quantities at the grid points. That this is a basic problem
especially for oceanogaphers is clearly visible in Figure 1 (taken from WOODS, 1985).
While meteorologists resolve at present the synoptic disturbances (mid-latitude de-
pressions), oceanographers cannot resolve the corresponding mesoscale eddies. Figure
1 also underlines that a large amount of parameterizations will always be needed in
the foreseeable future. For meteorologists parameterization also means detecting for
instance the possible development of a number of thunderstorms within a horizontal
grid cell and describing their effect on mean quantities on the grid just knowing
mean quantities at the grid.

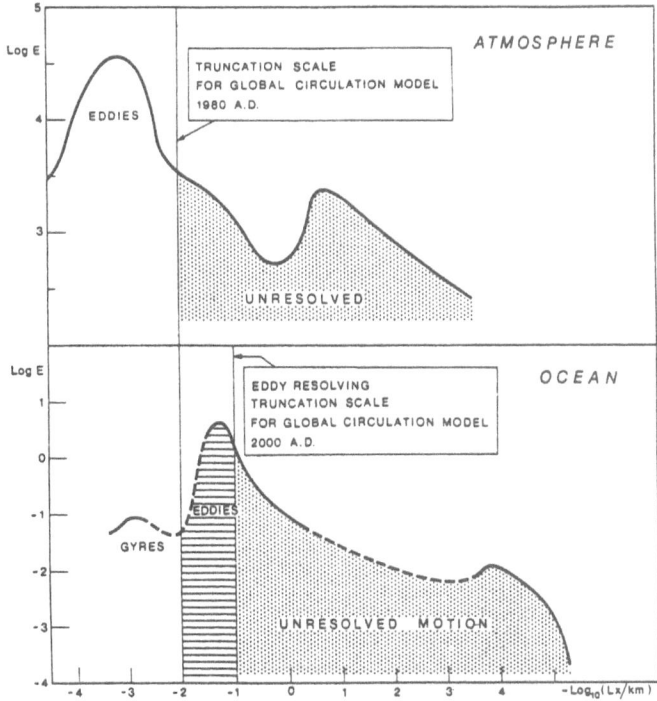

Figure 1: Typical spectra of motion in the ocean and atmosphere showing the energy
peaks associated with eddies. Existing computers permit a truncation scale
of 100 km. By the end of the century it should be possible to resolve the
ocean eddies.

3.3 Incomplete Equations

A third error, only sometimes tied to the computer capacity, is due to incomplete equations. This may be caused by the wish to avoid certain processes also described by the full equations, for instance sound waves, or just by lack of knowledge. Most three-dimensional climate models suffer strongly from this error as far as the biosphere contribution is involved.

3.4 Test of Climate Models

First of all, present climatic conditions should be reproduced by any climate model quite well. Since meteorologists maintain a global observing system in the atmosphere

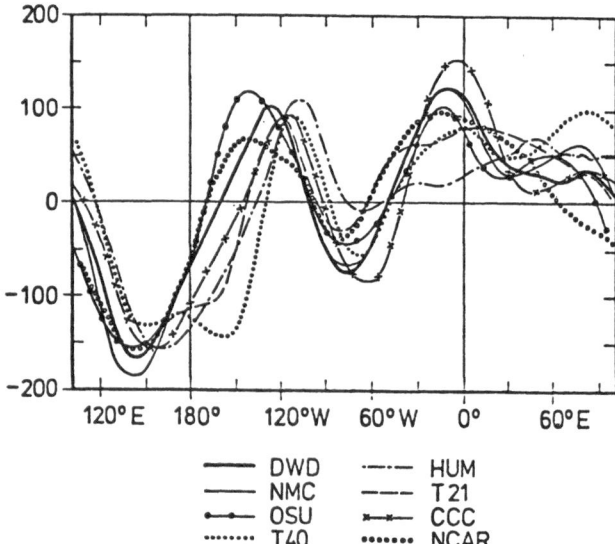

Figure 2: 30° - 60°N mean 500 hPa geopotential, January climatology (in geopotential meters) from six models and two observation analyses. DWD: German Met Service analyses from 1967 - 1985; NMC: US National Meteorological Center analyses from 1956 - 1966; T21 and T40: 2 versions of the European Centre for Medium Range Weather Forecast model resolving 21 or 40 waves along a latitude circle, HUM: Hamburg University; CCC: Canadian Climate Centre; NCAR: National Center for Atmospheric Research; OSU: Oregon State University.

the most advanced – because of continuous testing – climate models are those which emerged from weather forecast models. Figure 2 from STORCH et al, 1986, shows such a test. Six atmospheric general circulation models used as climate models, i.e. averaging over 30 days after having reached typical January conditions, are compared to two analyses of the 500 hPa pressure surface (~ 5 km height).

To attribute distinct errors to distinct models is nearly impossible, since they not only apply different grids but also contain or omit different parameterizations for subgrid scale processes. Nevertheless, the main troughs and ridges are given satisfactorily by all models. Please note that all six socalled climate models are atmospheric general circulation models with a very crude account of the ocean surface or surface layers only.

4. RESULTS FOR THE NATURAL SYSTEM

The distinction between natural and perturbed climate system is somewhat artificial, because mankind is part of the natural system and has since many years changed the surface albedo as well as evaporation and since industrialization also the composition of the atmosphere. Despite this dilemma we speak of climate model results for the natural system if no specific man-made changes have been imposed.

Although all modellers know that a climate model should include all relevant compartments a tested coupled ocean-atmosphere model is still not available. If atmosphere and ocean models are coupled the result is a model climate less realistic as the climate of single components driven by a fixed other component. The reason is found in small errors of momentum, heat and substance exchange at the boundary leading to large deviations in the coupled system through new feedbacks.

A first step towards tested coupling is shown in Figure 3 for the biggest and most persistent ocean surface temperature anomaly. Imposing the observed ocean surface temperature in the 20°S to 20°N area of the Pacific Ocean onto an atmospheric general circulation model the resulting pressure difference between Darwin, Australia, and Tahiti (a measure for the intensity of the El-Niño phenomenon) compares well with the observed (BIERCAMP et al, 1986).

A further step, a combination of the Hamburg University atmospheric general circulation model (emphasis on cloud formation) with an ocean deck layer model and a sea ice model confirms (see Fig. 4 and Table 1), that at least global and zonal means of a variety of climate parameters compare well with observations.

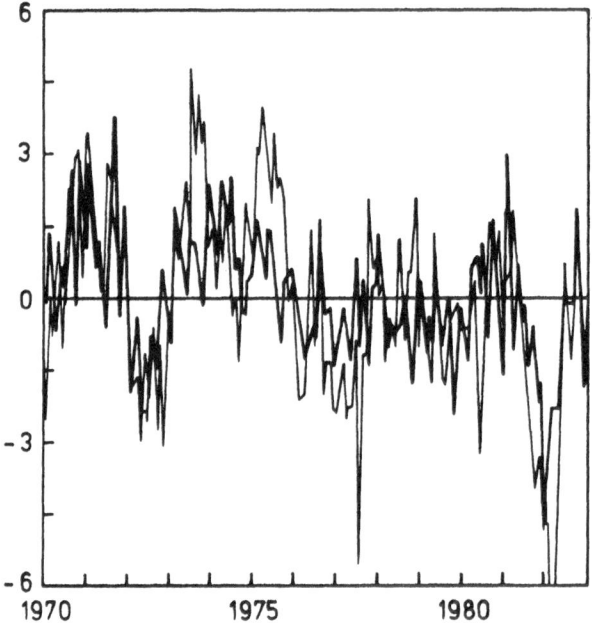

Figure 3: Monthly mean surface pressure difference Darwin minus Tahiti for 1970 to 1983; observed (—) and calculated (━).

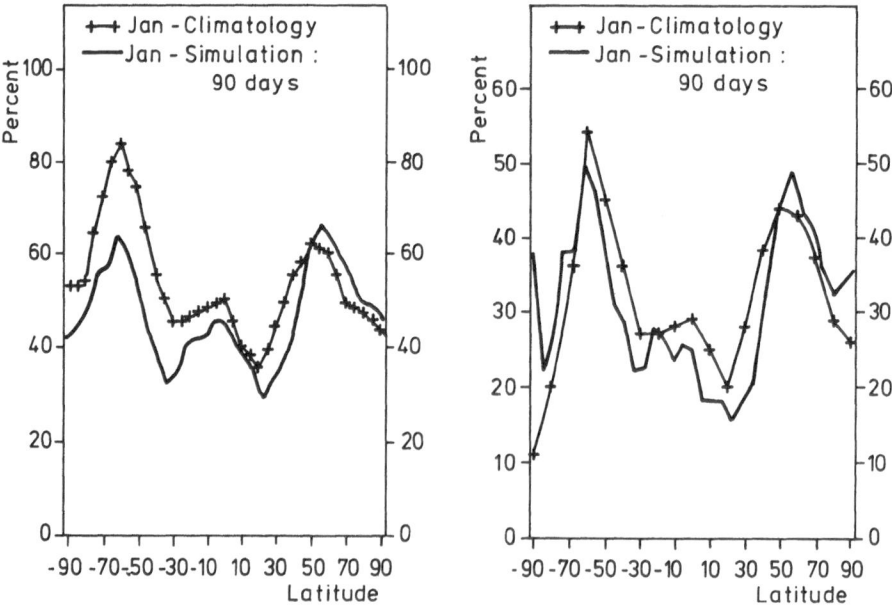

Figure 4: Simulated (——) versus observed (✱✱✱) zonal mean cloud cover for January as a function of latitude, left part: total cover, right part: low clouds.

Since some modelling groups are strongly engaged in the coupling of models we will soon have a variety of different results which hopefully will lead into a separation of that part of anomalies to be forecast and another part constituting climate 'noise' for instance within the time domain year-to-year variability.

Table 1: Calculated and observed global mean climate variables

Simulation Experiment	Surface temperature (°C)	Precipitation (mm/day)	Cloud Cover (%)	Planetary Albedo (%)	Sea Ice Cover (% of Earth Surface)
Fixed boundary conditions	·15.0	2.62	53.9	28.0	4.0
First 10 years coupled	14.7	2.59	53.3	27.9	4.11
Second 10 years coupled	14.6	2.49	53.2	27.8	4.67
Observation	14.9	2.74	52.1	~ 30	4.8

5. RESULTS FOR THE PERTURBED SYSTEM

A big part of the recent research effort in climatology is due to the observed regional and the anticipated global impact of man's activities on climate. Man's impact may be threefold:

- change of surface parameters like reflectivity, evaporation and roughness
- change of the atmospheric composition by emission of trace gases and aerosol particles
- waste heat.

Only the emission of trace gases into the atmosphere has been discussed long enough in order to give a somewhat consolidated picture; all the other man made changes are either not tackled in enough detail as for instance surface roughness change or are still not of global impact as waste heat.

5.1 Reaction to an Increased Trace Gas Content

Although the answer to the question: How will the climate system react to an increased atmospheric greenhouse effect? should be given by a time dependent coupled climate
model, all the public debate is still based on the stationary response of climate
models either uncoupled - as for general circulation models - or coupled for very
simplified one-dimensional models. There is no reason to discard an increased greenhouse effect by a number of gases with observed increased concentration (carbon dioxide (CO_2), methane (CH_4), chlorofluoromethanes (CFM's), dinitrogenoxide (N_2O) if
ordered according to importance at an anticipated increase). What is still a matter
of debate is the reaction of the global water cycle to this stimulus. Does slightly
increased temperature also increase the concentration of water vapour, the main
greenhouse gas? Will the type of clouds and cloud cover be changed? How does the
ocean circulation respond to changed atmospheric circulation? Surely, a part of these
questions is answered by three-dimensional atmospheric general circulation models and
nearly a consensus has been reached if stating: a doubling of CO_2-content will lead
to a mean global temperature increase between 1.5 and 4.5 K, with stronger warming in
high latitudes and lower and stronger warming in winter than in summer in low latitudes.

Since the regional pattern is not well simulated as shown during climate model tests
by comparison to present climate, mainly because of the poor space resolution (~ 500
km) of most climate models, only zonal mean changes are depicted in Figure 5.

Because the additional greenhouse effect of a well mixed gas leads to nearly identical results if compared to the CO_2 result, the combined action of the aforementioned
trace gases will not change the general temperature increase pattern. Recent estimates of future trace gas levels based on presently measured growth rates (RAMANATHAN
et al, 1985) have pointed to an equal importance of CO_2 and the other trace gas group
for the next 50 years.

When shall the doubling of CO_2 occur? In order to answer this question more than a
climate model is needed. The future energy consumption and industrial as well as
agricultural activity of mankind has been estimated frequently, but, if we remember
the 1972 forecasts for 1985, with no success. Therefore, Figure 6 displays a global
ocean model's reaction to a wide range of fossil fuel use patterns. The main result
of this global ocean model is: The CO_2 uptake of the ocean is inversely proportional
to CO_2 growth rates, or in other words, halving the growth rate more than doubles the
time needed for a distinct concentration increase.

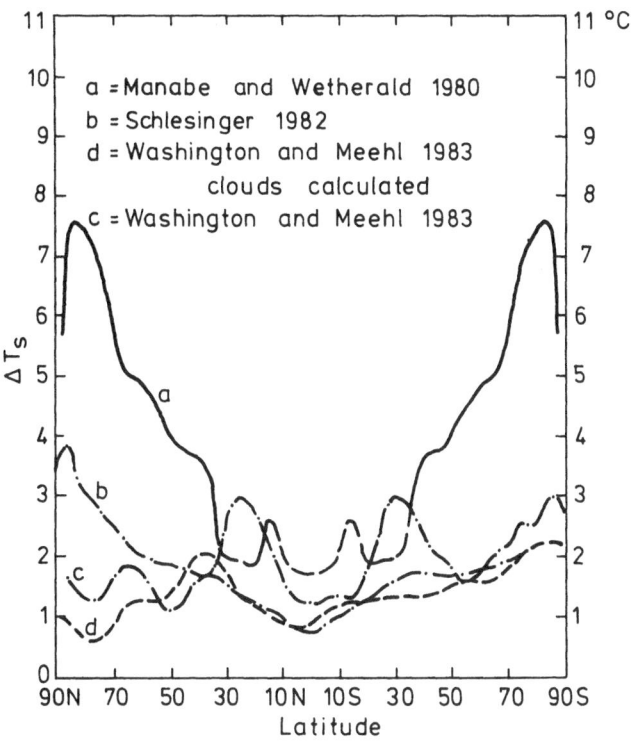

Figure 5: Zonal mean temperature increase due to a doubled CO_2 content as a function of latitude for different three dimensional models.

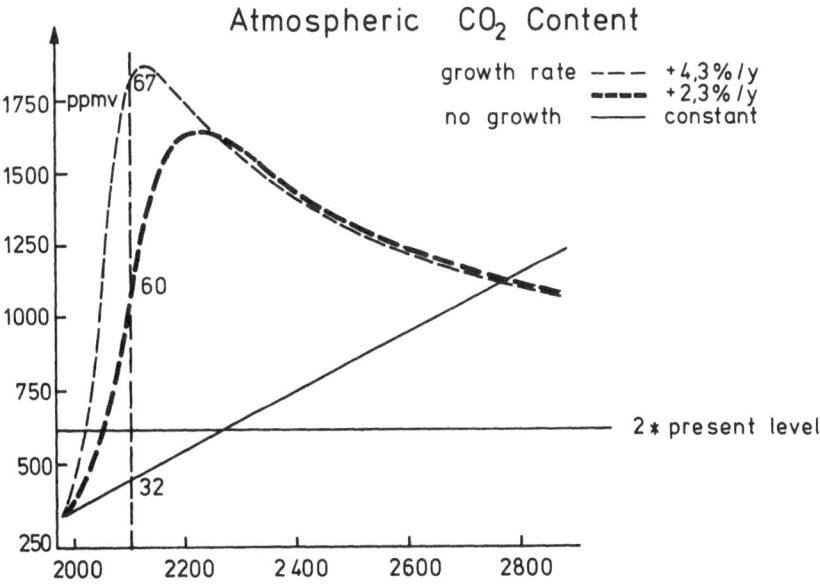

Figure 6: Atmospheric CO_2 content as a function of time for three fossil fuel use growth rates. The strongly different percentage remaining in the atmosphere shows the importance of the ocean uptake and its relation to growth rate.

5.2 Atmospheric Chemistry and Climate

The radiative transfer in the atmosphere is mainly determined by trace substances. Therefore, also gases or particles formed in the atmosphere from precursor substances not directly relevant to climate may be important. The outstanding substance in this respect is ozone, its concentration being the result of a large number of chemical reactions and horizontal as well as vertical transport. The simulation of future ozone levels in an atmosphere with a number of gases increasing in concentration is thus far more complicated than for CO_2. Up to now only results from one- or two-dimensional (vertical and latitudinal resolution) models of atmospheric chemistry have been published. Their main result is: Upper stratospheric ozone concentration decreases mainly depending on chlorofluoromethane growth rates (their photolytic destruction delivering the catalytic compounds for ozone destruction) while tropospheric ozone concentration increases mainly due to nitrous oxide emissions. Globally there is no compensation of both effects, the southern hemisphere expecting more total ozone reduction than the northern hemisphere.

5.3 Possible Effects of Aerosol Particles

For aerosol particles it is more difficult than for strongly varying gases to describe their behaviour. Additional important processes exist for aerosol particles:

- They are used as condensation nuclei
- they are formed through gas-to-particle conversion
- they change their size and composition due to coagulation
- they settle depending on size.

No three-dimensional model of aerosol transport is known to the author. However, a two-dimensional global tropospheric aerosol transport model coupled to a radiative transfer code in order to show radiation flux changes has been developed (GRASSL et al, 1984). It accounts for all the processes mentioned, uses anthropogenic particle input and allows for sulfuric acid formation from sulfur containing gases.

Due to the complexity of the processes involved and also due to the restriction to two dimensions the results should be interpreted very carefully. Nevertheless, as shown by NEWIGER (1985), (Fig. 7), the possible albedo change leads to net radiation flux changes of the same order of magnitude as for a doubling of CO_2 but of opposite

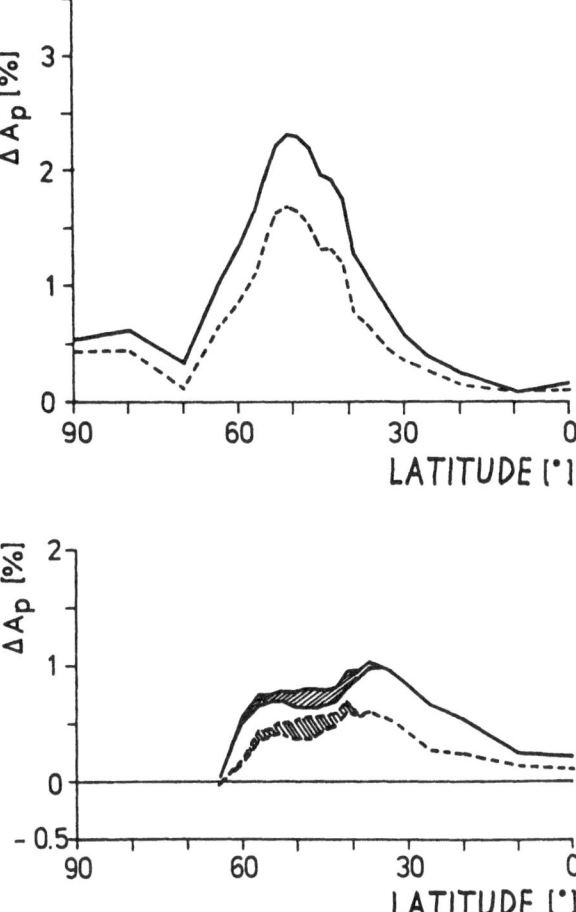

Figure 7: Albedo differences caused by present amounts of anthropogenic aerosol particles during summer (upper part) and during winter for particles with 20 %
soot content (---) or without soot. The shaded area shows the gas-to-particle conversion effect.

sign. The result is very sensitive to the soot content and the change in aerosol size
distribution. A big part of the albedo change is not due to aerosol particles in
cloudless areas alone, but is caused by the change in optical parameters of water
clouds, which was also taken into account.

The effect at the possible present anthropogenic aerosol load is too big to discard
it compared to the trace gas effect. However, consolidated results will not be reached easily and soon.

The influence of stratospheric aerosol particles on climate is no longer questioned. Since they are dominated by volcanic eruptions the anthropogenic influence should be small.

6. STRATEGY FOR A DETECTION OF CLIMATE CHANGE

The strong natural variability of climate has hampered a detection of the man-made contribution in climate parameter trends. Since even a stop of emissions, however, does not mean a stop of a possible climate change because of the long residence time of for instance chlorofluoromethanes and the slow reaction of climate components (deep ocean, ice sheets), there has to be searched for a strategy of rapid detection of a change.

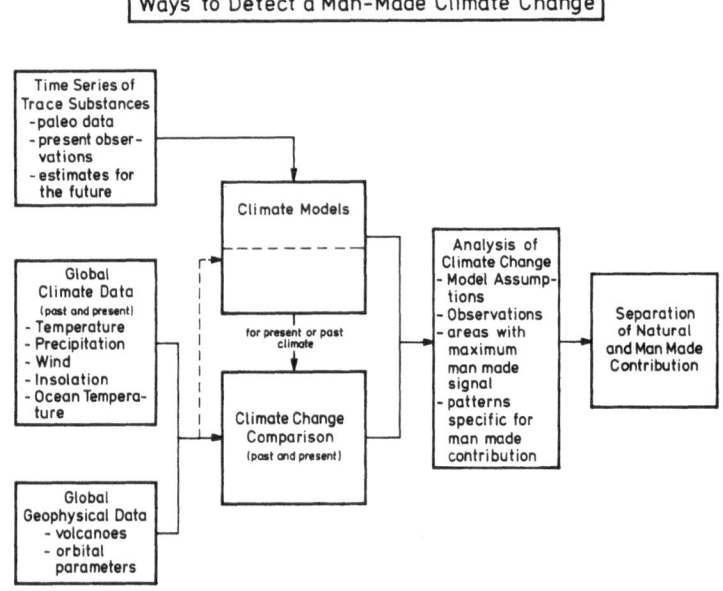

Figure 8: Ways to detect a man made climatic change (schematic)

The strategy proposed by many scientists is schematically displayed in Figure 8. The principle is a combination of observations with modelling. The observations must

- besides direct global climate observations - also include paleo data both for trace substances and surface parameters as well as geophysical data on volcanoes. Before model results are given for perturbations they have to be tested with present climate and/or paleo climate. After successful tests specific anticipated perturbations have to be applied to these models showing us areas of maximum relative man-made signal and specific patterns of a result of a perturbation. These hints then should be used to look with more intensity into observations in these areas or for these patterns.

The present uncertainty of the effect of mankind on climate should, in conclusion, not be the reason for no measures or precautions, instead we should give some probability also to the worst case.

REFERENCES

BIERCAMP, J.; LATIF, M.; von STORCH, H.; WRIGHT, P.B.; 1986: Preparational studies for coupling an oceanic and an atmospheric GCM; Research Activities in Atmospheric and Oceanic Modelling, CAS/JSC Working Group on Numerical Experimentation, Report # 9, 8.29 - 33, WMO, Geneva.

GRASSL, H.; LEVKOV, L.; NEWIGER, M.; REHKOPF, J.; 1984: Untersuchung des Einflusses anthropogen-bedingter Aerosolbildungen auf das Klima, Forschungsbericht 104 02 621, Umweltbundesamt, Berlin, 148 p.

MAIER-REIMER, E.; HASSELMANN, K; 1987: Transport and storage of CO_2 in the ocean - an inorganic ocean-circulation cycle model, submitted to Climate Dynamics 2, 63-90.

NEWIGER, M.; 1985: Einfluß anthropogener Aerosolteilchen auf den Strahlungshaushalt der Atmosphäre; Hamburger Geophysikalische Einzelschriften, Reihe A, Heft 73, 87 p.

RAMANATHAN, V.; CICERONE, R.J.; SINGH, H.B.; KIEHL, J.T.; 1985: Trace gas trends and their potential role in climate change, J. Geophys. Res. 90, D3, 5547 - 5566.

ROECKNER, E.; LÖWE, P.; BIERCAMP, J; 1986: Climate simulations with a simple coupled model; Research Activities in Atmospheric and Oceanic Modelling, CAS/JSC Working Group on Numerical Experimentation; Report 9, WMO, Geneva.

STORCH, H. v.; ROECKNER, E; CUBASCH, U.; 1985: Intercomparison of extended range January simulations with general circulation models: statistical assessment of ensemble properties; Beitr. Phys. Atm. 58, 477 - 497.

WMO, 1984: Scientific Plan for the World Climate Research Programme; WCRP Publ. Ser. No 2, Geneva.

WOODS, J.D.; 1985: The World Ocean Circulation Experiment, Nature 314, p. 501 - 511.

SENSITIVITY OF PRESENT-DAY CLIMATE TO ASTRONOMICAL FORCING

Ch. Tricot and A. Berger

Université Catholique de Louvain

Institut d'Astronomie et de Géophysique G. Lemaître

2 Chemin du Cyclotron

1348 Louvain-la-Neuve

1. Introduction

Many mechanisms have been suggested to explain the glacial-interglacial cycles during Quaternary (Berger, 1979b). The "astronomical" or "Milankovitch" theory involves variations of solar radiation available at the "top of the atmosphere" (called here extraterrestrial insolation) due to the secular perturbations of the Earth's orbit (e.g., Milankovitch, 1941; Berger et al., 1984; Berger and Tricot, 1986). Basically, the astronomical theory assumes that the surface air temperature is directly related to the insolation available at the Earth's surface for a completely transparent atmosphere and that the climate is sensitive to the changes in the distribution of that insolation among latitudes and seasons. The northern high latitudes are thought to be the most sensitive regions because of maximal continentality of these regions. Recent astronomical computations (Berger, 1984; Berger and Pestiaux, 1984) allow to claim that these variations of solar insolations are accurately known over the whole Quaternary and provide us with a well defined external forcing over this time period. The power spectra of the insolation variations display typical characteristic peaks near 40,000, 23,000 and 19,000 years, the first being associated with obliquity of the Earth's axis and the latter two with precession of the longitude of the perihelion (Berger, 1977). Since the pioneer paper of Hays et al. (1976), the astronomical theory has been given substantial support. That paper and others in the frequency (Pestiaux et al., 1987) and time (Imbrie and Imbrie, 1980) domains show that the Earth's orbital parameters play an important role in determining the succession of glacials and interglacials over the last million years.

However, there still remain difficulties in explaining how these changes in insolation could be sufficient to initiate or end glacial periods. Since a few years numerous climate models have tackled this problem but without definite conclusions. Simple climate models (e.g., energy balance models) cannot take into account in a correct way all relevant processes and feedbacks acting in the climate system. On the other hand, more complex numerical models (e.g., general circulation models) give results which are often less simple to analyse because of their inherent complexity. Also the use of 3-D models requires large amounts of computer time, which seriously limit their practical use for the study of long-term climatic changes including the transient response of the climate system.

In the present work, we focus on the astronomical forcing by examining the impact of the atmospheric attenuation upon the insolation changes. Indeed, the changes in the solar radiation absorbed within the atmosphere and by the surface can be thought physically more relevant for climate studies than the changes in the extra-terrestrial insolation itself. In section two the parameterization of the solar radiative transfer in the atmosphere used in our computations is described. In section three, some results for present-day climate are discussed. Finally, in section four, the long-term variations of the insolation received at the Earth's surface taking into account the present state of the atmosphere are presented and compared to the variations of the extraterrestrial insolation.

2. Parameterization of the shortwave radiative transfer

The factors that cause depletion of solar radiation in the atmosphere are absorption and scattering by gases, aerosols and clouds. A number of techniques to deal with the radiative transfer in an absorbing-scattering atmosphere are available (Lenoble, 1977). The delta-Eddington technique (Joseph et al., 1976) was used in our parameterization (called SOLAR) to handle the interactions between molecular absorption and scattering in a simple way.

For our application, a three-layer model of the atmosphere is assumed. The first layer extents from the surface to the pressure level of the cloud bottom, the second layer is filled up with an averaged cloud and the third layer extents from the pressure level of the cloud top to the top of the atmosphere. Three similar layers are defined for a clear sky. Ozone is assumed to be present only in the upper layer, while H_2O, CO_2, O_2 and aerosols are distributed within each layer.

In order to be used in the delta-Eddington approximation, the scattering parameters for a homogeneous layer containing more than one scattering or absorbing component are combined to give a set of effective parameters for the layer. The effective optical thickness, single scattering albedo and asymmetry factor are given, respectively, by

$$\delta_e = \Sigma \, \delta_i \qquad\qquad (1.1)$$

$$\omega_e = \Sigma \, \delta_{sc,i} / \delta_e \qquad\qquad (1.2)$$

$$g_e = \Sigma \, \delta_{sc,i} \, g_i / \delta_{sc} \qquad\qquad (1.3)$$

where $\delta_{sc,i} = \omega_i \delta_i$ is the optical thickness due to scattering by the atmospheric component i,

$$\delta_{sc} = \omega_e \delta_e ,$$

δ_i, ω_i and g_i are respectively the optical thickness, the single scattering albedo and the asymmetry factor of the atmospheric component i.

According to Fouquart (1986), one single spectral interval (0.25-4µm) will be considered. Parameterizations of the spectrally averaged optical thickness for the Rayleigh scattering and of the spectrally averaged transmittances for H_2O, CO_2, O_2 and O_3 are used in each layer. For a gaseous absorber, the optical depth in a layer j is given by

$$\Delta\delta_j = - \mu_j \ln [Tr_{j+1} / Tr_j] \qquad\qquad (2)$$

where Tr_j is the transmittance due to the gaseous amount encountered by the solar radiation from the top of the atmosphere to layer j, before entering it,

Tr_{j+1} is similarly defined but includes the gaseous amount encountered in the layer,

μ_j is the cosine of the effective zenith angle of the solar radiation going through the layer j.

This definition of the gaseous optical depth for a layer follows the method proposed by Braslau and Dave (1973). In our scheme, downward radiation in each layer is treated as a collimated radiation with the same zenith angle as the incident radiation at the top of the atmosphere, in the case of a clear sky. For cloudy sky, the same assumption is made for the upper layer, while the downward radiation into and below the cloud is assumed to be diffuse, the same being assumed for the upward radiation in both cloudy and clear conditions. For diffuse radiation, the angular integration to obtain fluxes is avoided by using the diffusivity approximation : the radiation is assumed to be collimated with a zenith angle chosen here equal to 53 degrees.

The delta-Eddington approximation gives the reflection and transmission of a single homogeneous layer with fixed values of δ_e, ω_e and g_e. To combine the three layers, we use the ascending method presented by Fouquart and Bonnel (1980, see also Fouquart, 1986). Starting from the layer near the surface and accounting for the combined reflectance of the underlying layers, the reflectance R_{j+1}^+ at level j+1 and the transmittance T_j^- of layer j are given by :

$$R_{j+1}^+ = R_j^+ + R^- \, T_j \, T_j^{\ast\ast} \, / \, (1 - R_j^{\ast\ast} R^-) \tag{3}$$

and $\quad T_j^- = T_j \, / \, (1 - R_j^{\ast\ast} R^-)$ $\qquad\qquad\qquad\qquad\qquad\qquad$ (4)

where R_j and T_j are the reflectance and transmittance of the isolated layer j from above and given by the delta-Eddington parameterization,

$\qquad R^-$ is the combined reflectivity of the underlying layers,

$\qquad R_j^{\ast\ast}$ and $T_j^{\ast\ast}$ are the reflectance and transmittance of the isolated layer j from below and given by the delta-Eddington parameterization.

The denominator of Eqs (3) and (4) accounts for the multiple reflections between the layers. This term is especially important when the reflection between layers is large which will typically be the case when a thick cloud overlies a bright surface (e.g., a snow covered surface). The differentiation in Eqs (3) and (4) between the reflectance and transmittance from above and below takes into account that the reflectance and transmittance of a homogeneous isolated layer depend on the incidence direction of the solar radiation. For the layer near the surface, R^- is equal to the surface reflectivity or surface albedo.

Assuming a downward numbering of the layers, the downward and upward fluxes at level j are given, respectively, by :

$$F_j^\downarrow = F_{top}^\downarrow \, \prod_{k=1}^{j-1} T_k^-$$

$$F_j^\uparrow = F_j^\downarrow \, R_j^+$$

where $F_{top}^\downarrow$ is the downward extraterrestrial solar radiation.

Finally, partial cloudiness is treated by calculating the fluxes separately for a clear and a completely overcast sky and then weighting linearly the respective fluxes by the clear and cloudy fractions :

$$F_j^\downarrow = (1-n) \, (F_j^\downarrow)_{clear} + n (F_j^\downarrow)_{cloudy}$$

$$F_j^\uparrow = (1-n) \ (F_j^\uparrow)_{clear} + n(F_j^\uparrow)_{cloudy}$$

with n the fractional cloud amount.

3. Present-day insolation pattern at the surface

This section provides the annual variation of mid-month daily insolation at the Earth's surface for the present-day atmospheric conditions. The computations were undertaken with the solar parameterization (SOLAR) introduced in the previous section and were performed for each latitude from 90°N to 90°S by step of 10° and for each month.

To obtain "mid-month" insolation, a constant increment of the true solar longitude λ is used starting with $\lambda = 0$ at vernal equinox. The mid-month values are thus defined by $\Delta\lambda = 30°$ and they are located now around the 20th of each month. Because the length of the astronomical seasons is secularly variable (Berger, 1978), these mid-month values are not related to a fixed calendar date in the past. However, in all the computations presented here, we will consider mid-month insolation as representative of monthly mean insolation.

Figure 1.a shows firstly the distribution of the daily incident solar radiation at the Earth's surface for a transparent atmosphere. The classical analytical expressions of this extraterrestrial insolation are given in Berger (1978) and depend only on the duration of sunlight, the latitude, the solar declination, the distance between the Earth and the Sun and the value of the solar constant (taken here equal to 1368 W m^{-2} following Willson et al. (1981)). Since the Sun is presently closest to the Earth in January (during northern hemisphere winter), the distribution of solar energy is slightly asymmetrical and the maximum radiation received in the southern hemisphere is greater than that received in the northern hemisphere. The maximum insolation occurs at summer or winter solstice at either pole owing to the long solar day (24 hours). The annual variations are maximum at poles and decrease regularly to the equator.

To estimate the atmospheric attenuation and compute the solar energy incident at the Earth's surface, some atmospheric and surface characteristics must be known. In our computations, we have used the zonally and monthly averaged fields of surface temperature and humidity published by Oort (1983). The vertical profiles are deduced from the surface values with Rennick's (1977) and Smith's (1966) parameterizations for temperature and humidity, respectively. An effective single cloud is assumed with monthly cloud amount from Berlyand and Strokina (1980). For each latitude, the

137

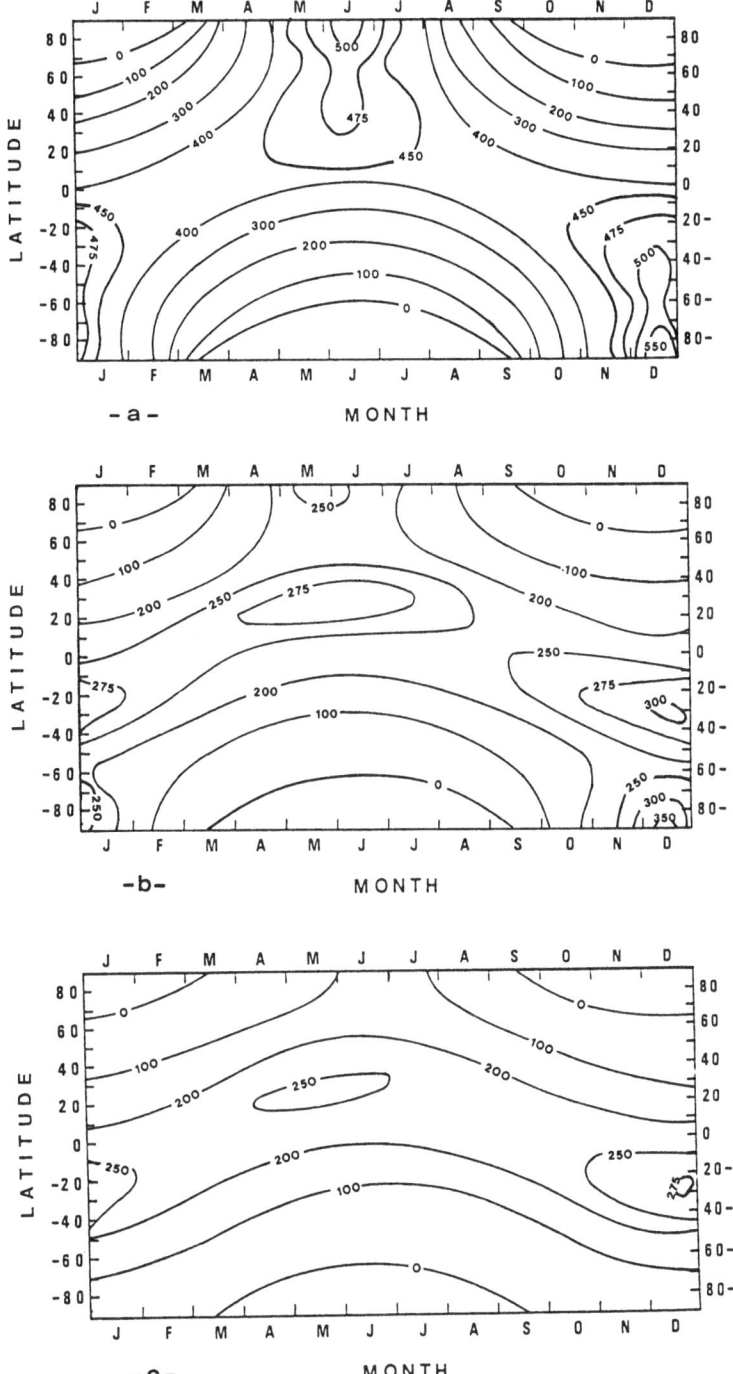

Figure 1. The daily variation of present insolation as a function of latitude and day of year in units of Wm^{-2} (a) at the top of the atmosphere, (b) incident at the surface, (c) absorbed at the surface.

cloud position and optical thickness are kept fixed through the year. The values of these parameters are adopted from Chou et al. (1981), except that their values for cloud optical thickness were diminished by approximately 2 between 20°N and 20°S to reduce the differences between computed and observed planetary albedos. For cloud, the asymmetry factor g is taken equal to 0.85 and the single scattering albedo is related to the fixed optical thickness following Fouquart and Bonnel (1980). The total ozone amount is computed following the parameterization given by Van Heuklon (1979) and the CO_2 concentration is fixed to 330 ppmv. Three kinds of aerosol models were considered : maritime, continental and unperturbed stratospheric, following WMO (1986). D. Tanre (personal communication) provided us with the spectrally averaged scattering parameters for each aerosol model and the optical thickness at 0.55 μ m were taken from a WCRP report (WMO, 1981). The zonally averaged vertical profile for aerosols is calculated by weighting the vertical profiles above ocean and continent (except at 80°S and 90°S, where the aerosols were considered as maritime aerosols in the planetary boundary layer). Finally, the surface albedos were interpolated from data given by Robock (1980).

Contrary to the case of a transparent atmosphere, daily insolation W at the Earth's surface, when the atmospheric attenuation is taken into account, requires a daily integral to be solved numerically. This integral can be written as follows :

$$W = S_0 \left(\frac{r_m}{r} \right)^2 \int_{-H_0}^{H_0} Tr(\cos z) \cos z \, dH \qquad (5)$$

where S_0 is the solar constant,

H_0 is the hour angle of the sunrise,

r_m is the mean distance from the Earth to the Sun,

r is the actual distance between the Earth and the Sun,

z   is the solar zenith angle,

$Tr(\cos z)$ is the atmospheric transmission, function of $\cos z$ and computed by the solar model.

To save computer time, the daily insolation given by Eq.(5) can be approximated by using a daily mean atmospheric transmission which allows Eq.(5) to be written :

$$W = S_0 \left(\frac{r_m}{r} \right)^2 \overline{\cos z \, Tr} \qquad (6)$$

$$\text{with } \overline{\cos z} = \int_{-H_0}^{H_0} \cos z \ dH \ / \ \int_{-H_0}^{H_0} dH$$

and $\overline{Tr}$ is computed using an effective daily mean cosine of the zenith angle, $(\cos z)^*$.

Our SOLAR parameterization was compared with a version of the more sophisticated SUNRAY model developed by Fouquart and Bonnel (1980). Both solar models use the same Rayleigh parameterization and gaseous transmittances. Multiple scattering is solved in the two models by the delta-Eddington scheme. However, in SUNRAY, the interactions between scattering and molecular band absorption are taken into account by an extension of the photon path length distribution method, which deals with the scattering and absorption processes separately. SUNRAY also uses a higher vertical resolution (between 10 and 15 layers).

In order to illustrate this comparison, Table 1 gives, for some selected latitudes, the mid-month daily insolation absorbed at the Earth's surface for January and July as computed with Eq.(5) using SUNRAY (row A, daily integration) and SOLAR (row B, daily integration) and with Eq.(6) using SOLAR (row C). The comparison of rows A and B shows that SOLAR overestimates slightly and systematically the daily insolation at the surface. The closest agreement between rows A and C was obtained by using $(\cos z)^*$ given by Eq.(7), except for the highest latitudes where the differences are similar to these between rows A and B :

$$(\cos z)^* = 0.75 \ \overline{\cos z} + 0.25 \ \text{cosmax} \tag{7}$$

where cosmax is the cosine of the daily minimum value of z, i.e., its value at solar noon, given by $\cos (\phi - \delta)$ with ϕ the latitude and δ the solar declination. Similar results were obtained for the other months.

Figures 1.b and 1.c give the distribution of, respectively, the present daily incident and absorbed solar radiation at the surface computed with SOLAR and using Eq.(6) with $(\cos z)^*$ given by Eq.(7). The difference between these two figures is due to the surface albedo which defines the fraction of the incident solar radiation reflected by the surface. The impact of this factor is particularly visible in summer high latitudes. Furthermore, the difference between the extraterrestrial insolation pattern (shown in Fig. 1.a) and the pattern for the incident solar radiation at the surface (Fig. 1.b) is related to the atmospheric transmission which decreases from low to high latitudes mainly as a result of the increase with latitude of the

averaged solar zenith angle and of cloudiness. In consequence, the maximal insolation at the surface is located now in the tropics during summer with a secondary
maximum at the summer pole in the northern hemisphere. In the southern hemisphere,
the maximal insolation is yet above Antarctica during summer with a secondary maximum in the tropics.

Table 1. Mid-month daily insolation absorbed at the Earth's surface for present
January and July at some selected latitudes. Results of row A are obtained with
a version of SUNRAY (Fouquart and Bonnel, 1980) integrated through the day.
Results of row B are obtained with the present SOLAR parameterization integrated
through the day and results of row C are given by SOLAR using the effective
daily averaged cosine of the zenith angle as defined by Eq.(7). The insolation
is given in Wm^{-2}.

Latitude	A SUNRAY		B SOLAR		C SOLAR with $(\cos z)^*$	
	January	July	January	July	January	July
70	0	145	0	149	0	141
50	28	201	29	205	29	201
30	116	246	118	250	117	248
0	221	211	224	214	221	211
-30	270	113	274	115	271	113
-50	229	34	232	35	227	35
-70	121	0	126	0	119	0

To test further the reliability of our computations for present-day conditions,
we have compared these results to the zonal averages of the annual mean absorbed
solar radiation as computed by some other authors (Stephens et al., 1981; Peng et
al., 1982; Ou and Liou, 1984) and with the zonal averages of planetary albedos as
deduced from satellite observations (Stephens et al., 1981; Hartmann et al., 1986).
These comparisons, not presented here, show that our model calculations agree generally well with all these other theoretical studies and observations.

4. Past insolations at the Earth's surface

In order to estimate past climatic variations, the changes in the absorbed solar
radiation by the atmosphere and at the surface can be thought physically more rele-

vant than the changes in the extraterrestrial radiation. An accurate estimate of these variations over the last million years requires the surface characteristics and the atmospheric constituents to be known over this period. A well known example concerns the surface albedo, which is directly related to the ice sheet growths and decays and the sea ice extent, mainly around the Antarctica. Qualitative estimates of the changes of some parameters begin to be available for some particular regions and times of the Quaternary period (e.g., Jouzel et al., 1982; Lorius et al., 1984; Kutzbach and Street-Perrott, 1985). However, continuous and global time series recording the variations of these parameters are far for being yet available. Nevertheless, some sensitivity studies could be undertaken as a first step to the simulation of past insolations and climates.

In this section, the variations of the absorbed and incident solar radiation at the Earth's surface are computed for a period extending from 200 kyr BP (Before Present) to present, keeping all atmospheric and surface characteristics equal to their present-day values. Thus such computations must be viewed only as a sensitivity study of the present-day climate to the well known variations of the astronomical forcing. A similar work was undertaken by Ohmura et al. (1984) but using a simplified form of the global atmospheric transmission and considering only the solar radiation incident (not absorbed) at the surface.

According to the Milankovitch astronomical theory of paleoclimates (Milankovitch, 1941), ice ages are initiated when cool summers in high latitudes can lead to the persistence of the snow field over the continental areas all years long. Therefore, we will only discuss in this study the variations of insolation for July (summer for northern hemisphere) and January (summer for southern hemisphere). For this application, the approximation given by Eq.(7) has been further validated at 125 kyr BP and 115 kyr BP again by comparing for these periods the results given by SUNRAY and SOLAR. During these two particular periods, July experienced the largest positive (at 125 kyr BP) and the largest negative (at 115 kyr BP) deviations from present-day conditions for the northern hemisphere over the last 200 kyrs (Berger, 1979a). Table 2 gives the mid-month absorbed solar radiation at the Earth's surface for some particular latitudes and for January and July as computed by SOLAR with Eq.(7) and by SUNRAY with a daily integration. As for present-day computations, both models agree very well which reinforces our confidence to use our parameterization for the whole period considered here.

Table 2. Mid-month daily insolation absorbed at the Earth's surface in January and
July for 125 kyr BP and 115 kyr BP at some selected latitudes. SUNRAY is based
on the model of Fouquart and Bonnel (1980) integrated through the day while
SOLAR is the present parameterization used with the effective daily averaged
cosine of zenith angle as defined by Eq.(7). The insolation is given in Wm^{-2}.

	125 KBP				115 KBP			
Latitude	SUNRAY		SOLAR		SUNRAY		SOLAR	
	January	July	January	July	January	July	January	July
70	0	165	0	161	0	132	0	131
50	25	227	25	226	32	188	33	187
30	103	277	103	278	126	233	127	234
0	198	236	197	236	235	203	234	203
-30	243	125	243	125	282	111	283	111
-50	206	37	204	38	235	35	230	35
-70	110	0	108	0	122	0	121	0

Figures 2.a and 2.b show the long-term variations of the deviations from today
values of solar radiation for January. Each figure represents 100 kyrs. Figures 3.a
and 3.b show similar variations for July. In these figures, the top panel represents
the extraterrestrial solar energy, the middle one the incident solar energy at the
Earth's surface and the bottom one the absorbed solar energy at the Earth's surface.
As discussed by Berger (1979a), it is especially interesting to note the large positive deviations for the July insolation at the top of the atmosphere in high latitudes for 10 kyr BP (40 to 50 Wm^{-2}), for 85 kyr BP (40 Wm^{-2}), for 103 kyr BP (50 Wm^{-2})
and for 125 kyr BP (60 Wm^{-2}). All these deviations represent more than 10 per cent
of the insolation received today at these latitudes and their timing corresponds
closely to the timing of the warm and cold stages of the last 130 kyrs.

The variations of the incident solar radiation at the surface (middle panels)
show similar pattern but the absolute value of the variations are reduced by a factor 2 at low latitudes and up to around 4 at high latitudes. In fact, the atmospheric attenuation reduces the absolute values of the variations as expected, but
not the relative ones. For the variations of the absorbed solar radiation at the
surface (bottom panels), the largest deviations are no longer found in high latitudes but in tropical and middle latitudes. The increase of the surface albedo with
latitude reduces further the absolute values of the insolation and therefore its
absolute variations. Because of the high summer surface albedo for latitudes south-
ward of 65°S, this feature for the absorbed insolation is especially well pronounced
in Fig. 2.a and 2.b.

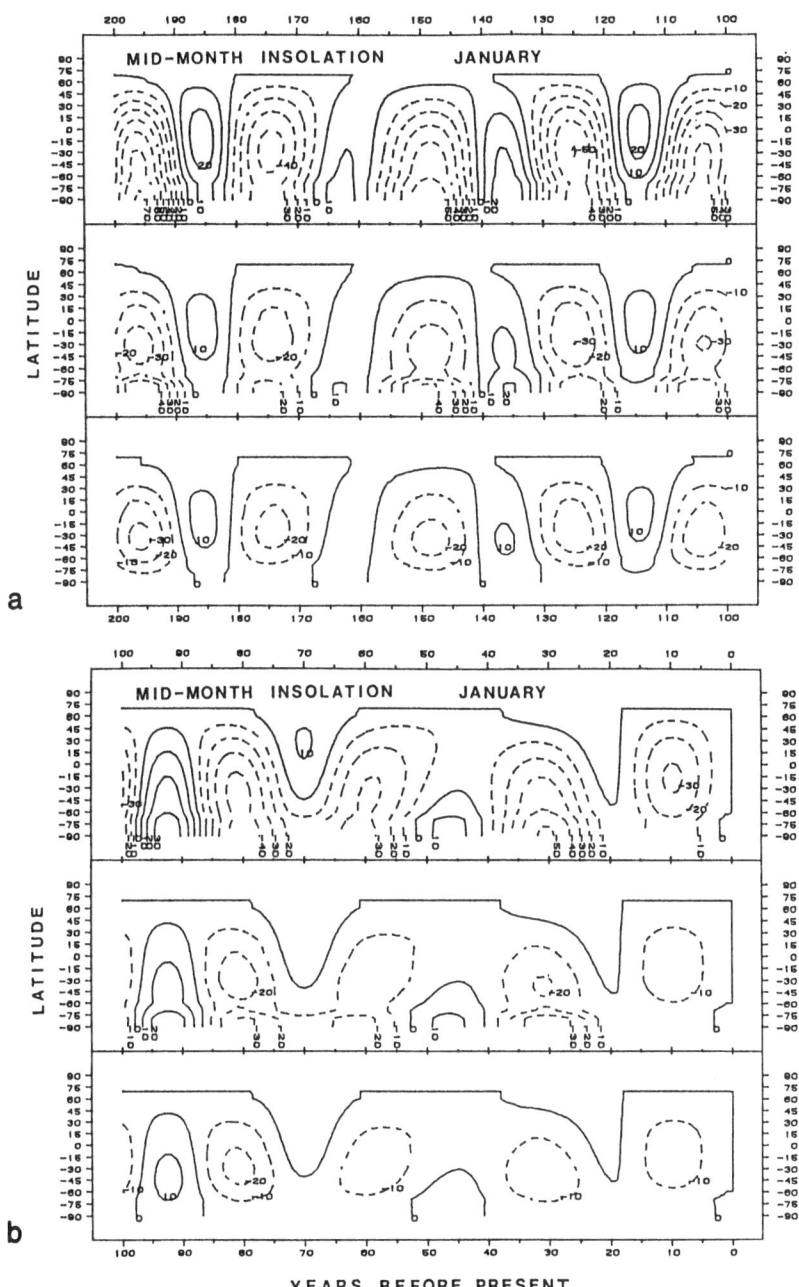

Figures 2.a-b. Long-term variations of the deviations (from their present values) of mid-month daily insolations for January. These values are given here in Wm^{-2} and for periods extending from 100 to 200 kyr BP (part a) and from present to 100 kyr BP (part b). For each time period, the top panel is for the insolation at the top of the atmosphere, the middle one for the incident insolation at the surface and the bottom one for the absorbed insolation at the surface. The solid lines are for the positive deviations (insolation higher than today) and the dashed lines characteristize the insolation below their present values.

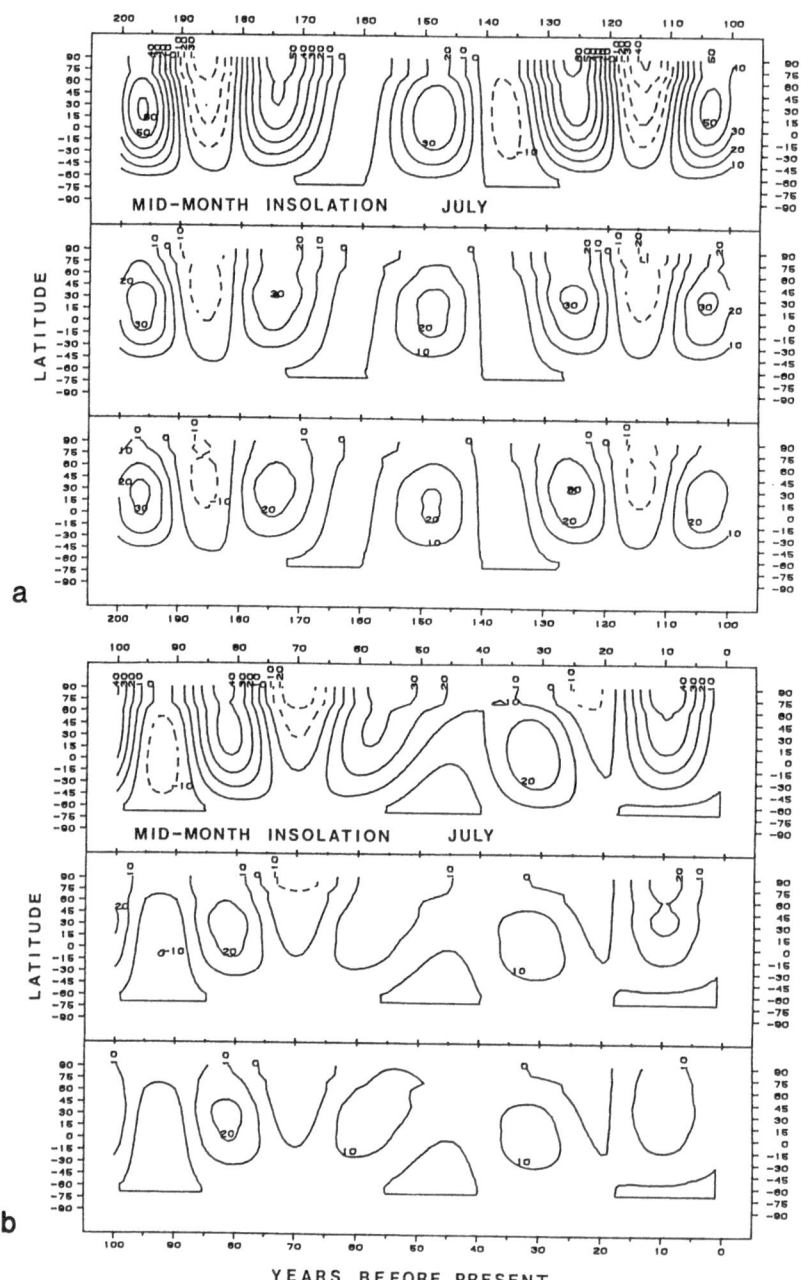

Figures 3.a-b. Long-term variations of the deviations (from their present values) of mid-month daily insolations for July. These values are given here in Wm^{-2} and for periods extending from 100 to 200 kyr BP (part a) and from present to 100 kyr BP (part b). For each time period, the top panel is for the insolation at the top of the atmosphere, the middle one for the incident insolation at the surface and the bottom one for the absorbed insolation at the surface. The solid lines are for the positive deviations (insolation higher than today) and the dashed lines characteristize the insolation below their present values.

To link insolation to climate, the variations of the latitudinal gradient of insolation is as important (if not more) as the local variations of insolation. Instead of using a local gradient, as done by Ohmura et al. (1984), a large-scale gradient between tropical and high latitudes has been defined by calculating the difference between insolation at 30° and 70° in each hemisphere. Figure 4a shows the long-term variations of January mid-month insolation over the last 200 kyrs for 30°N and 30°S. Three curves as in figures 2 and 3 (respectively extraterrestrial insolation, incident at and absorbed by the surface) are again drawn for each latitude. Figure 4.b gives similar curves for 70°N and 70°S and Figure 4.c gives the variations of the large-scale gradient of insolation, obtained by substracting Fig. 4.b from Fig. 4.a. Finally, Figures 5.a, 5.b and 5.c are similar figures but for July.

For 30°N and 30°S, Fig. 4.a and 5.a again illustrate the similar decrease of the variations of the absolute insolations incident and absorbed at the surface, when the atmospheric attenuation is taken into account. These curves are almost identical because of the small surface albedo in the tropical regions. On the contrary, Fig. 4.b and 5.b show the importance of the larger surface albedo at 70°N (around 25 per cent in July) and 70°S (around 55 per cent in January). As mentioned above, the difference between incident and absorbed radiation is especially visible for 70°S in January because of the relatively larger surface albedo. In the winter hemisphere, the variations of insolation in high latitudes (70°N in Fig. 4.b and 70°S in Fig. 5.b) are almost continuously equal to zero because the insolation is equal or near zero over the whole period at these latitudes. This peculiarity helps to interpret partially the variations of the large-scale gradient of insolation given by Fig. 4.c and 5.c. For the winter hemisphere (northern for Fig. 4.c and southern for Fig. 5.c) the variation of the gradient is given by the variation of the insolation at 30° alone (compare the three upper curves of Fig. 4.a with the three upper curves of Fig. 4.c and the three lower curves of Fig. 5.a with the three lower curves of Fig. 5.c).

For the summer hemisphere, the variation of the gradient is not so easy to interpret. Interestingly, for the north-south gradients in summer hemispheres (Fig. 4.c SH and 5.c NH), the characteristic frequencies of the absorbed insolation at the surface are no more those of the extra-terrestrial insolation. From spectral analysis, Berger and Pestiaux (1984) have indeed shown that extraterrestrial mid-month insolation at 30°N and 30°S presents mainly peaks around 19 and 23 kyrs, while mid-month insolation at high latitudes shows in summer an additional peak around 41 kyrs (such features can be deduced from analysis of Fig. 4.a, 4.b, 5.a and 5.b). As far as the variation of the latitudinal gradient is concerned, Fig. 4.c and 5.c show that extraterrestrial insolation is characterized by a periodicity of about 40 kyrs, whereas the absorbed insolation exhibits a higher frequency which corresponds to a period in the range of 23 kyrs. The incident insolation at the surface follows the

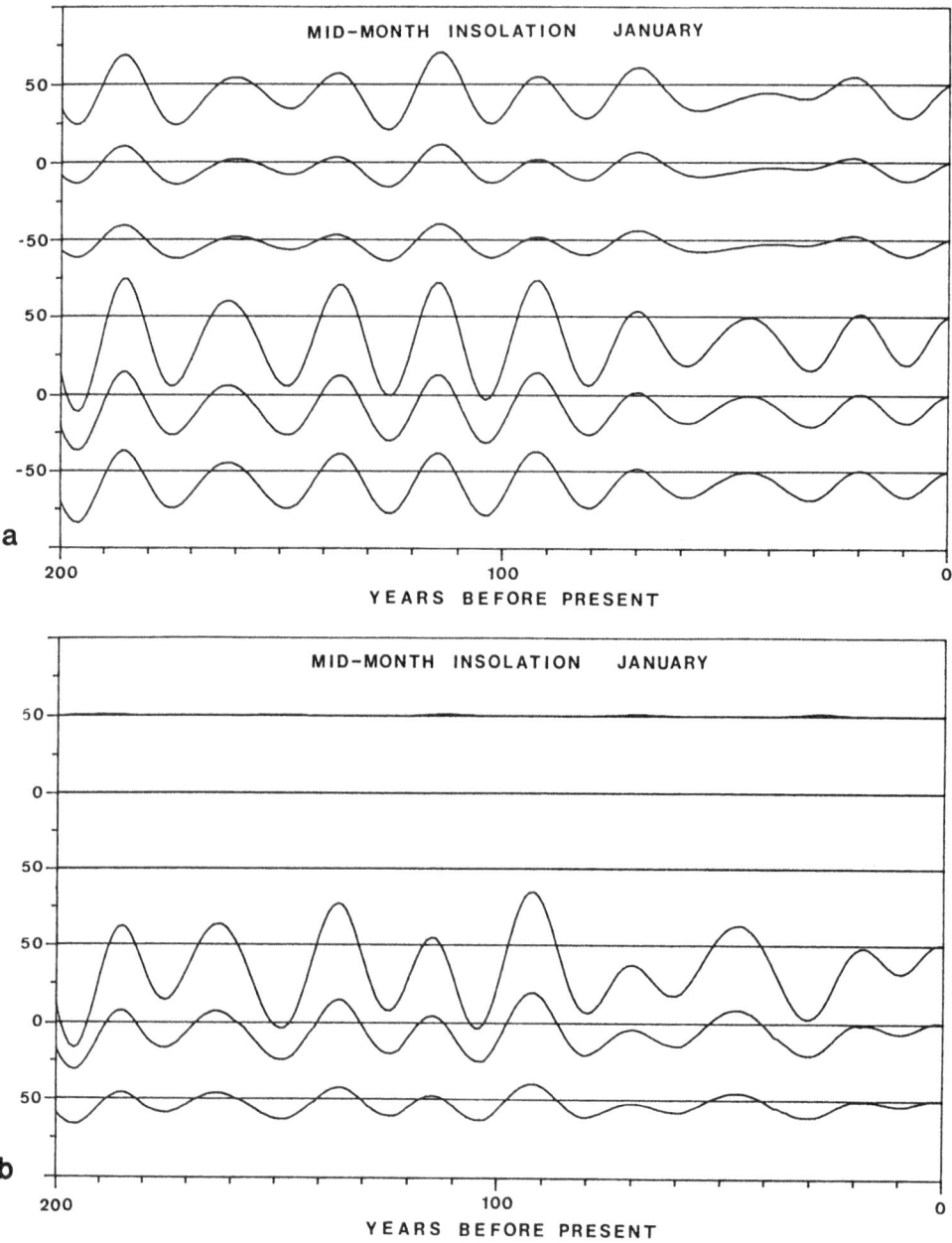

Figures 4.a-b. Long-term variations of January mid-month insolation over the last 200 kyr, in Wm^{-2}, for 30°N and 30°S (part a) and for 70°N and 70°S (part b). The left scale gives the deviation from the present values. For each latitude in both hemispheres, the upper curve gives the variations for the solar radiation at the top of the atmosphere, the middle curve the variations for the incident radiation at the Earth's surface and the lower curve the variations for the absorbed radiation at the Earth's surface.

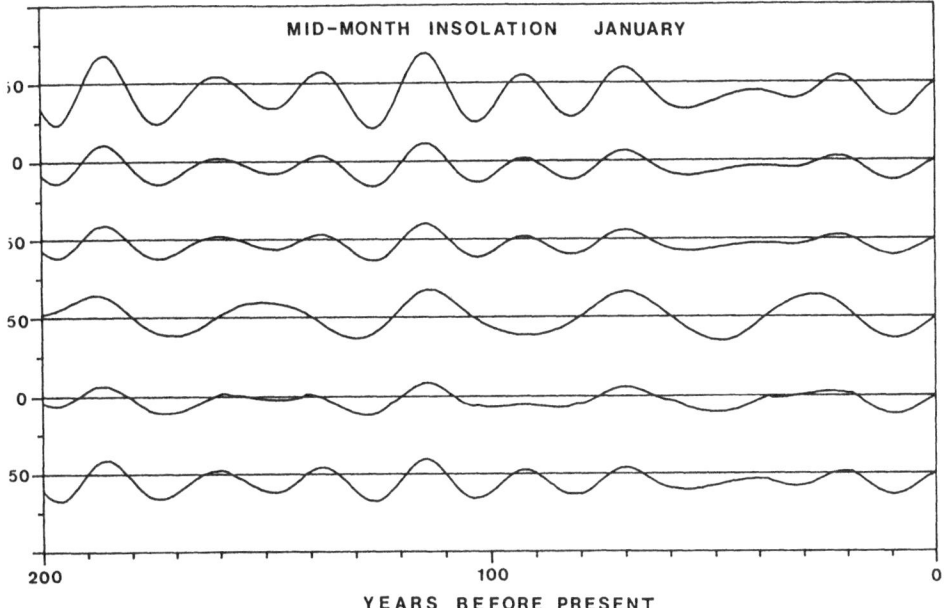

Figure 4.c. Long-term variations of January mid-month large-scale gradient of insolation over the last 200 kyr, in Wm^{-2}, for the northern hemisphere (the three
upper curves), and the southern hemisphere (the three lower curves). The left
scale gives the deviation from the present values of the gradient defined as the
difference between insolation at 30° and 70°. For each hemisphere, the upper
curve gives the variations for the solar radiation at the top of the atmosphere,
the middle curve the variations for the incident radiation at the Earth's
surface and the lower curve the variations for the absorbed radiation at the
Earth's surface.

same frequency behaviour than the absorbed radiation, at least for the northern
hemisphere summer (Fig. 5.c) for this similarity is less clear for the southern
hemisphere summer (Fig. 4.c). If the extraterrestrial summer insolation curves of
Fig. 4.a (30°S) and 4.b (70°S) or Fig. 5.a (30°N) and 5.b (70°N) are compared (it
means January southern hemisphere - lower panel of Fig. 4.a and 4.b - and July
northern hemisphere - upper panel of Fig. 5.a and 5.b), the periodicity of their
difference (given in 4.c lower panel and 5.c upper panel, respectively) can be explained by noting that the successive maximal deviations (in absolute value) from
today are sometimes larger at 30° than at 70°, and sometimes it is the opposite. The
change in time of the location of these relative maxima induces a periodicity of
about 40 kyrs in the variations of the latitudinal gradient. On the contrary, for
the insolations at the surface, the successive maxima are, most of the time, located
at 30° and the large-scale gradient exhibits the same periodicity than the insolation at that latitude. These larger deviations from today values in the tropical
insolations are due to the larger atmospheric attenuation in high latitudes which
reduces more the available insolation (and therefore its variations) at these latitudes than in the tropics.

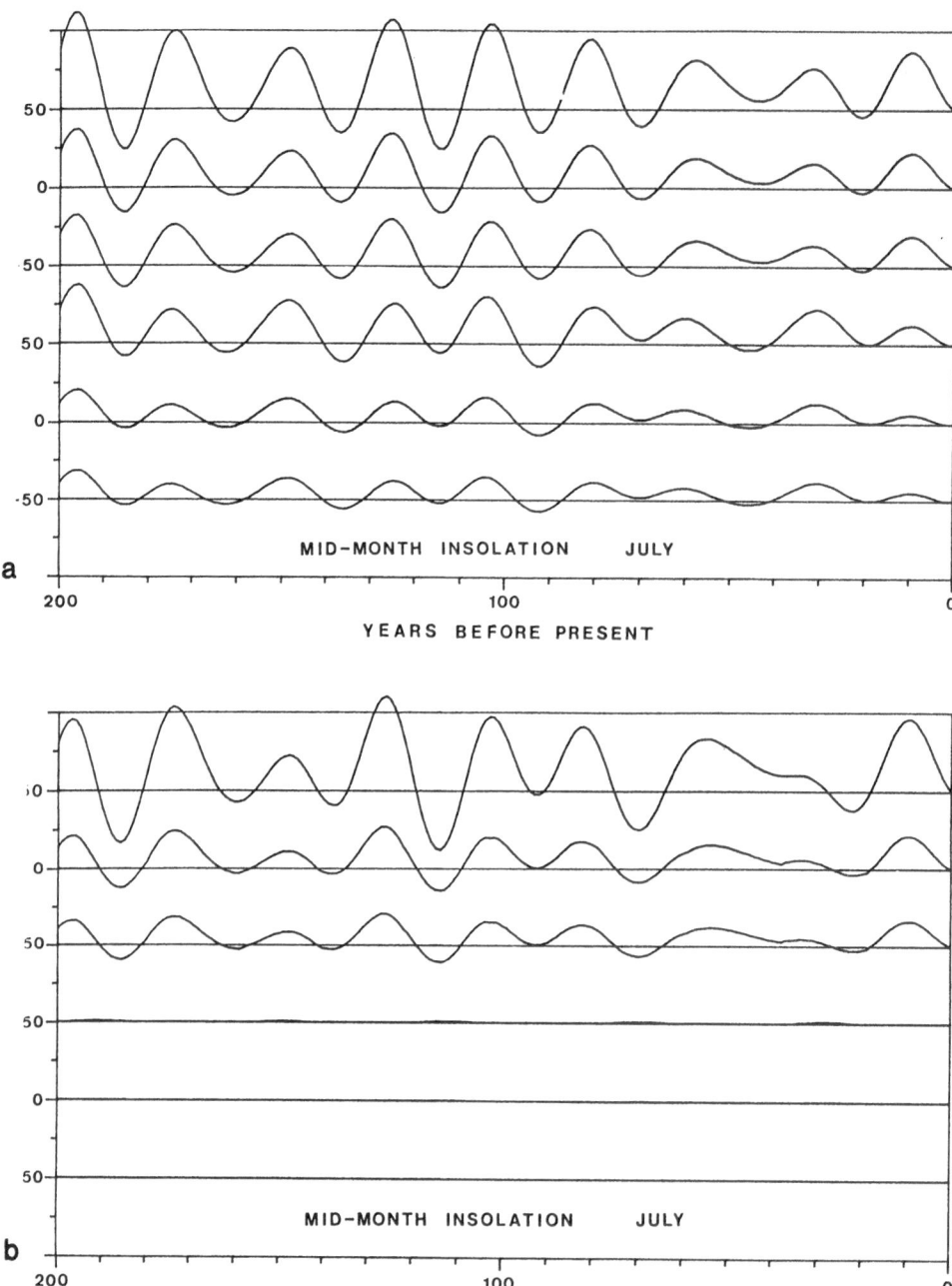

Figures 5a-b. Long-term variations of July mid-month insolation over the last 200 kyr, in Wm^{-2}, for 30°N and 30°S (part a) and for 70°N and 70°S (part b). The left scale gives the deviation from the present values. For each latitude in both hemispheres, the upper curve gives the variations for the solar radiation at the top of the atmosphere, the middle curve the variations for the incident radiation at the Earth's surface and the lower curve the variations for the absorbed radiation at the Earth's surface.

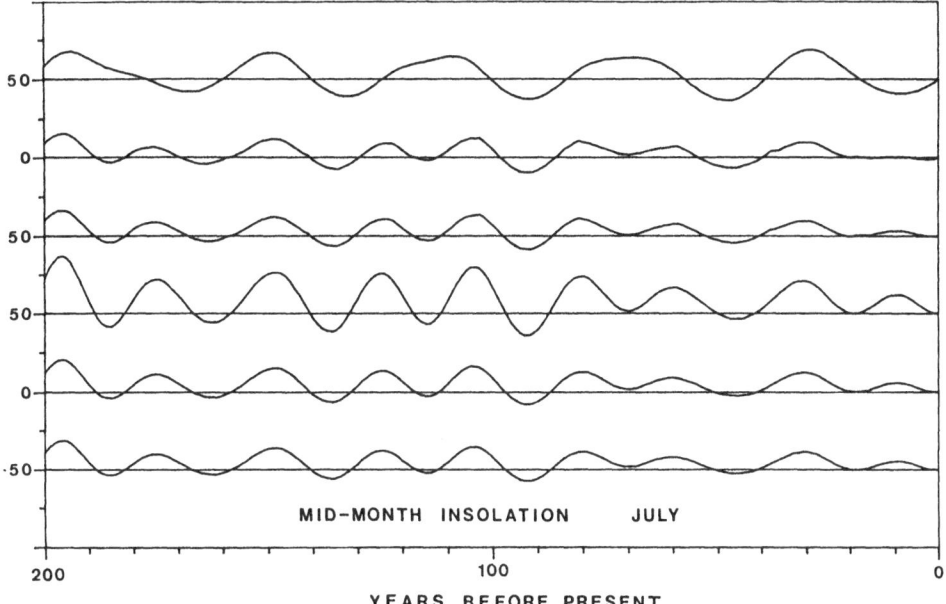

Figure 5.c. Long-term variations of July mid-month large-scale gradient of insolation over the last 200 kyr, in Wm^{-2}, for the northern hemisphere (the three upper curves) and the southern hemisphere (the three lower curves). The left scale
gives the deviation from the present values of the gradient defined as the difference between insolation at 30° and 70°. For each hemisphere, the upper curve
gives the variations for the solar radiation at the top of the atmosphere, the
middle curve the variations for the incident radiation at the Earth's surface
and the lower curve the variations for the absorbed radiation at the Earth's
surface.

5. Conclusions

According to the astronomical theory of paleoclimates, the long-term variation
in the geometry of the Earth's orbit is the fundamental cause of the succession of
Pleistocene ice ages. Accurate values for the variations of these orbital elements
and related monthly insolations are now available for the last 2 to 3 million years.
Even if recent climatic models, both qualitative and quantitative, show that the
orbital parameters have modulated the climate during the whole Quaternary (and will
probably continue to do so assuming no human interferences), the exact mechanisms
which link insolation variations to climate variations at the astronomical frequencies are not yet totally known. Both simple models, which would reproduce the dyna-
mic behaviour of climatic changes through time, and more sophisticated models which
allow, in particular, to test the validity of the first for selected dates, must be
developed further.

In this paper, an analysis of the impacts of the insolation forcing on the insolation available at the Earth's surface has been made by comparing, in the time and frequency domains, variations of the extraterrestrial radiation to variations of the incident and absorbed radiations at the Earth's surface. Considering the potential importance of the insolation during summer (which could prevent or allow snow melting), results for July for northern hemisphere and January for southern hemisphere have been stressed. The atmospheric attenuation essentially reduces the absolute variations of the incident solar radiation at the Earth's surface as compared to the variations of the extraterrestrial radiation. Over the last 200 kyr, these two kinds of insolation generally present maximal variations in high latitudes in relation with the variation of the obliquity. On the contrary, the absorbed radiation at the Earth's surface has always maximal variations in tropical and middle latitudes related to the increase of the surface albedo with latitude.

Finally, in the summer hemisphere, the large-scale gradient of insolation between the tropics and the polar regions shows deviations from its present-day value which characteristic frequencies depend upon the type of insolation considered : (i) for the extraterrestrial insolation, the main frequency of the variations of the large-scale latitudinal gradient is about 40 kyrs, whereas (ii) for the incident and mainly the absorbed insolation at the surface the large-scale gradient shows, in addition, quasi-periodicity of about 23 kyrs; this difference is related to the atmospheric attenuation which reduces more strongly the variations of insolation at the surface in high latitudes than in tropics.

References

Berger, A. 1977 : Support for the astronomical theory of climatic change. Nature, 269, 44-45.
Berger, A. 1978 : Long-term variations of daily insolation and Quaternary climatic changes. Journal of Atmospheric Science, 35(12), 23622367.
Berger, A. 1979a : Insolation signatures of Quaternary climatic changes. Il Nuovo Cimento, series 1, 2, 63-87.
Berger, A. 1979b : Spectrum of climatic variations and their causal mechanisms. Geophysical Surveys, 3(4), 351-402.
Berger, A. 1984 : Accuracy and frequency stability of the Earth's orbital elements during the Quaternary. In : A. Berger, J. Imbrie, J. Hays, G. Kukla, B. Saltzman (Eds) : Milankovitch and Climate, pp. 3-40, D. Reidel Publ. Company, Dordrecht, Holland.
Berger, A., Imbrie, J., Hays, J., Kukla, G., Saltzman, B. (Eds) 1984 : Milankovitch and Climate, D. Reidel Publ. Company, Dordrecht, Holland.
Berger, A. and Pestiaux, P. 1984 : Accuracy and stability of the Quaternary terrestrial insolation. In : A. Berger et al. (Eds) : Milankovitch and Climate, pp. 83-112, D. Reidel Publ. Company, Dordrecht, Holland.
Berger, A., and Tricot, Ch. 1986 : Global climatic changes and astronomical theory of paleoclimates. In : A. Cazenave (Ed.) : Earth Rotation : Solved and Unsolved Problems, pp. 111-129, D. Reidel Publ. Company, Dordrecht, Holland.
Berlyand, T.G., and Strokina, L.A. 1980 : Zonal cloud distribution on the Earth. Meteor. Gidrol., 3, 15-23.

Braslau, N., and Dave, J.V. 1973 : Effect of aerosols on the transfer of solar energy through realistic model atmospheres. Part I : Nonabsorbing aerosols. J. Appl. Meteor., 12, 601-615.

Chou, S.H., Curran, R.J., and Ohring, G. 1981 : The effects of surface evaporation parameterization on climate sensitivity to solar constant variations. J. Atmos. Sci., 38, 931-938.

Fouquart, Y. 1986 : Radiative transfer in climate modeling. In : M. Schlesinger (Ed.) : Proceedings of the NATO-ASI on Physically-Based modeling and simulation of climate and climatic change, Erice, 11-23 May 1986, to be published by D. Reidel Publ. Company, Dordrecht, Holland.

Fouquart, Y., and Bonnel, B. 1980 : Computations of solar heating of the Earth's Atmosphere : a new parameterization. Beitr. Phys. Atmos., 53, 35-62.

Hays, J.D., Imbrie, J., and Shackleton, N.J. 1976 : Variations in the Earth's orbit : pacemaker of the ice ages. Science, 194, 1121-1132.

Hartmann, D.L., Ramanathan, V., Berroir, A., and Hunt, G.E. 1986 : Earth Radiation Budget-Data and climate research. Rev. Geophys., 24, 439-468.

Imbrie, J. and Imbrie, J.Z. 1980. Modeling the climatic response to orbital variations. Science, 207, 943-953.

Joseph, J.H., Wiscombe, W.J., and Weinman, J.A. 1976 : The delta-Eddington approximation for radiative flux transfer. J. Atmos. Sci., 33, 2452-2459.

Jouzel, J., Merlivat, L., and Lorius, C. 1982 : Deuterium excess in an East Antarctic ice core suggests higher relative humidity at the oceanic surface during the last glacial maximum. Nature, 299, 688-691.

Kutzbach, J.E., and Street-Perrott, F.A. 1985 : Milankovitch forcing of fluctuations in the level of tropical lakes from 18 to 0 kyr BP. Nature, 317, 130-134.

Lenoble, J. 1977 : Standard procedures to compute atmospheric radiative transfer in scattering atmospheres. Proc. IAMAP Radiation Commission, NCAR, Boulder, 125pp.

Lorius, C., Raynaud, D., Petit, J.R., Jouzel, J., and Merlivat, L. 1984: Late glacial maximum-Holocene atmospheric and ice thickness changes from Antarctic ice core studies. Annals of Glaciol., 5, 88-94.

Milankovitch, M.M. 1941 : Kanon der Erdbestrahlung. Beograd. Köninglich Serbische Akademie. 484pp. (English translation by israël program for Scientific Translation and published for the U.S. Department of Commerce and the National Science Foundation).

Ohmura, A., Blatter, H., and Funk, M. 1984 : Latitudinal variation of seasonal solar radiation for the period 200,000 years BP to 20,000 AP. In : G. Fiocco (Ed.) : IRS 84 : Current problems in Atmospheric Radiation, Proceedings of the International Radiation Symposium, Perugia, Italy, 21-28 August 1984, A. Deepak Publ., 338-341.

Oort, A.H. 1983 : Global atmospheric circulation statistics, 1958-1973. NOAA Professional Paper 14, U.S. Gov. Printing Office, Washington, 180pp.

Ou, S.S., and Liou, K.N. 1984 : A two-dimensional radiation turbulence climate model. I. : Sensitivity to Cirrus radiative properties. J. Atmos. Sci., 41, 2289-2309.

Peng, L., Chou, M.D. and Arking, A. 1982 : Climate studies with a multilayer energy balance model. Part I : Model description and sensitivity to the solar constant. J. Atmos. Sci.., 39, 2639-2656.

Pestiaux, P., Duplessy, J.Cl., van der Mersch, I., and Berger, A. 1987 : Paleoclimatic variability at frequencies ranging from 1 cycle per 10,000 years to 1 cycle per 1,000 years : evidence for nonlinear behavior of the climate system. To be published in Climatic Change.

Rennick, M.A. 1977 : The parameterization of tropospheric lapse rates in terms of surface temperature. J. Atmos. Sci., 34, 854-862.

Robock, A. 1980 : The seasonal cycle of snow cover, sea ice and surface albedo. Mon. Wea. Rev., 108, 267-285.

Smith, W.L. 1966 : Note on the relationship between total precipitable water and surface dew point. J. Appl. Meteor., 5, 726-727.

Stephens, G.L., Campbell, G.G. and Vonder Haar, T.H. 1981 : Earth Radiation Budgets. J. Geophys. Res., 86, 9739-9760.

Van Heuklon T.K. 1979 : Estimating atmospheric ozone for solar radiation models. Solar Energy, 22, 63-68.

Willson, R.C., S., Hanssen, M., Hadson, H.S., and Chapman, G.A. 1981 : Observations of solar irradiance variability. Science, 211, 700-702.

152

WMO 1981 : Aerosols and Climate. report of the Meeting of JSC experts, Geneva, 27-31 October 1980. WCRP report, WCP-12, 60p.

WMO 1986 : A preliminary cloudless standard atmosphere for radiation computation. WCRP report, WCP-112, WMO/TD-n°24.

CAUSES AND EFFECTS OF NATURAL CO_2 VARIATIONS DURING THE GLACIAL-INTERGLACIAL CYCLES

Ulrich Siegenthaler
Physics Institute, University of Bern
Bern, Switzerland

1. Observations

Although carbon dioxide is a minor atmospheric constituent, its volume concentration being only 0.03 percent, it plays a very funda-mental role in nature - for the earth's climate via the greenhouse ef-fect, for vegetation as a basic substance for photosynthesis. Varia-tions of its concentration may therefore have great impact on the en-vironmental conditions on earth, and it was an exciting finding when several years ago analyses of samples of old polar ice indicated that during glacial time, the atmospheric CO_2 level was reduced to about 70 % of its recent pre-industrial level. This observation immediately raised two questions: (1) what was the cause of the glacial-postgla-cial CO_2 concentration shift, and (2) what are the implications for the ice age climate. In the following, I shall first discuss shortly the observational evidence for the glacial-interglacial CO_2 varia-tions, then the present-day ideas about possible causes and finally the estimated effects on global climate.

The principle of the ice core measurements is simple. During the formation of ice by sintering of cold firn, air bubbles are enclosed which have the composition of atmospheric air. Extraction of the trap-ped air and analysis in the laboratory (usually by means either of gas chromatography or of infrared laser spectroscopy) permit to determine the concentration of CO_2 - and of other trace gases - in these samples of the ancient atmosphere. However, meaningful results can be obtained only if several conditions are fulfilled. First, melting must never have occurred in the ice, because CO_2 is highly soluble in water, so that melting of ice and refreezing leads to layers enriched in CO_2. For this reason, only ice from sites with a mean annual temperature below about -20°C, i.e. from polar regions, can be used for such analyses; at warmer sites, melting events do occur in summer. Second, the determination of the concentration in the laboratory requires special skill. For avoiding CO_2 dissolution effects, air extraction in the laboratory is done by crushing the ice at e.g. -20°C. One kg of polar ice typically contains 100 ml of air or 30 µl of CO_2, so that an ice sample of typically 10 g to 500 g mass contains rather little CO_2. In addition, CO_2 tends to behave indecently in laboratory systems. Adsorption and desorption from walls occur, especially in the presence of water vapour, and CO_2 is preferentially transported, compared to other air components, by water vapour in extraction systems [Zumbrunn et al., 1982].

Figure 1 shows CO_2 concentrations measured on the deep ice core from Dye 3, Greenland [Stauffer and Oeschger, 1985]. At the transition between ice from postglacial and glacial time, at ca. 1810 m depth, the values decrease from about 300 ppm to 200 ppm. If this shift would have been observed in only one ice core, one might suspect that it is caused by some interaction between CO_2 and the surrounding ice, e.g. due to alkaline ice from glacial time. However, a similar pattern of concentration change has been observed in six different ice cores so far, two from Greenland and four from Antarctica, representing very different temperatures and ice properties [Delmas et al., 1980, Neftel et al., 1982, Stauffer et al., 1985, Barnola et al., 1987]. Thus, the observed concentrations most likely do represent atmospheric values. The records show that the atmospheric CO_2 concentration during the later part of the Würm glaciation were near 200 ppm and rose between roughly 14,000 and 10,000 B.P. to the Holocene value of ca. 280 ppm. The concentration history during the Holocene is not well established, partly because of brittle ice quality in the corresponding depth intervals, but does not seem to have undergone large variations. (The values above 300 ppm at Dye 3, Figure 1, may be due to summer melt events.) The measurements of Barnola et al. [1987] on the Vostok ice core indicate low values during the whole Würm glacial, similar CO_2 concentrations as during the Holocene in the last interglacial, ca. 120,000 yr B.P., and again low values during the penultimate glaciation. Thus, it seems that atmospheric CO_2 was generally low during glacial periods and relatively high during interglacials.

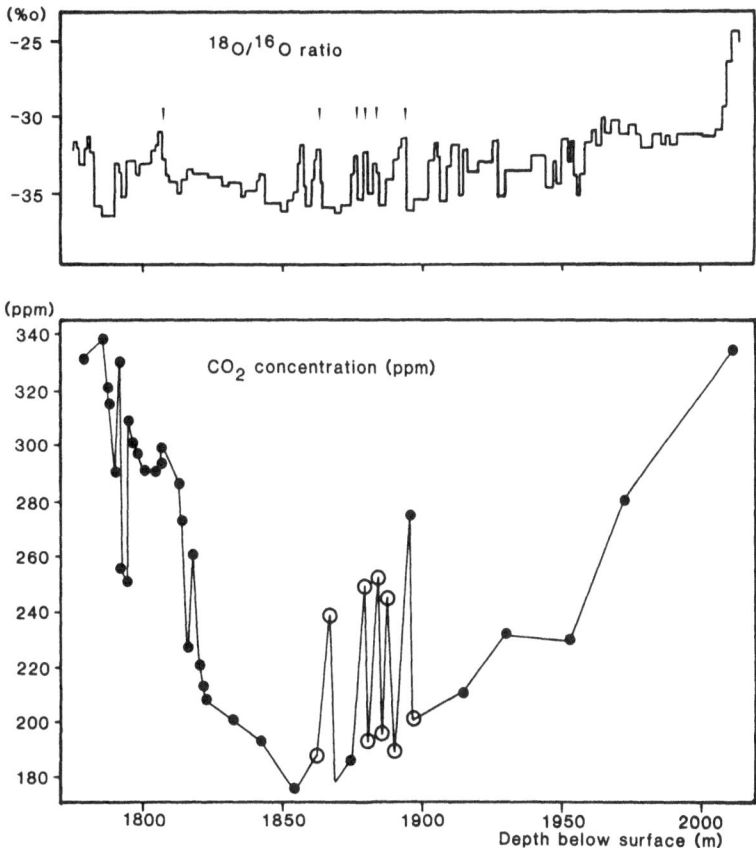

Figure 1: CO_2 concentrations measured on the deep ice core from Dye 3, Greenland [Stauffer and Oeschger, 1985]; top: $^{18}O/^{16}O$ ratios, indicating climate change (measured by Dansgaard and colleagues). Time scale: the $^{18}O/^{16}O$ shift at ca. 1782 m depth occurred about 10,000 years B.P., that at ca. 1806 m depth about 13,000 years B.P., the interval 1860-1885 m depth (arrows) to 30,000-40,000 years B.P.

A problem of special interest is the rate of concentration change with time. According to the available data the full shift from 200 to 280 ppm at the end of the last glaciation occurred within about 4000 years. In the Dye 3 core, intervals of relatively high CO_2, near 250 ppm, have been found during glacial time, in the period of about 30,000 to 40,000 yr B.P. (open circles in Figure 1), and detailed measurements indicated that the CO_2 concentration shifted, together with

$\delta^{18}O$ and other ice properties, abruptly at these depths (Figure 2) [Stauffer et al., 1984]. If an undisturbed ice stratigraphy is assumed, then the changes of CO_2 and, simultaneously, of $\delta^{18}O$ occurred within a few centuries or less. Such rapid variations would have considerable implications for their possible causes. However, thorough and dense measurements on the ice core from Byrd Station, Antarctica, do not show similar rapid variations [Neftel et al., 1987]. The situation is, therefore, not clear; and the possibility must be considered that the rapid transitions observed in the Dye 3 ice core are due to disturbed stratification (e.g. "sandwiching" of older between younger ice) of the glacial ice in Greenland.

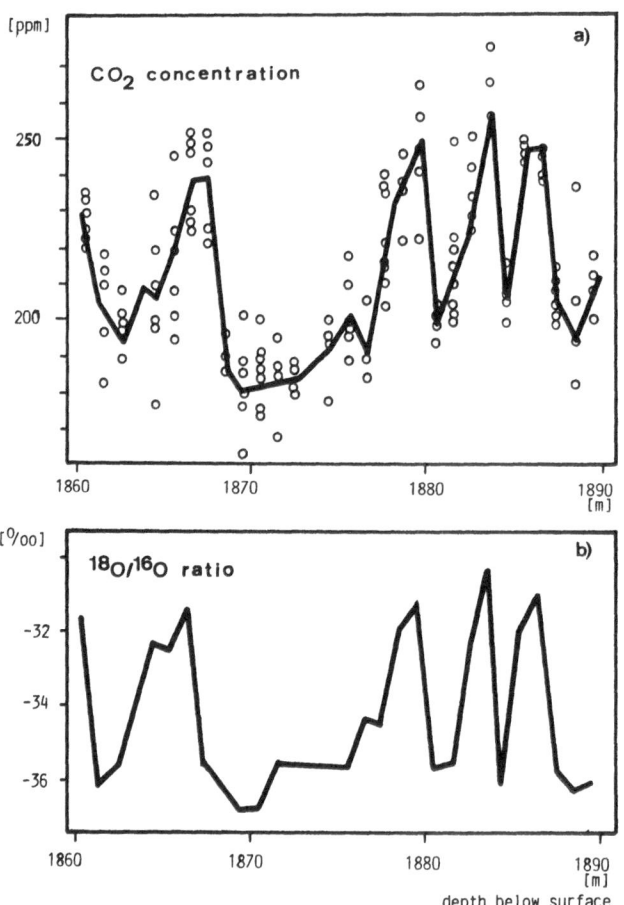

<u>Figure 2:</u> CO_2 concentration and $^{18}O/^{16}O$ ratio on a section of the ice core from Dye 3, Greenland, of ca. 30,000 - 40,000 years B.P. Note the synchronicity of the rapid changes in the two parameters [Stauffer et al., 1984].

2. Possible causes of the CO_2 variations

2.1   The oceanic carbon cycle

The CO_2 variations must be due to a redistribution of carbon between its major global reservoirs - atmosphere, ocean, land biosphere and ocean sediments. Volcanism and tectonic processes need not be considered here, since they are slow and only become important on time scales > 10^6 years. The most important reservoir is the ocean, containing about 60 times as much carbon as the atmosphere. The atmospheric CO_2 concentration is directly regulated by the chemical composition of sea surface water. It is therefore necessary to consider the carbonate chemistry of sea water.

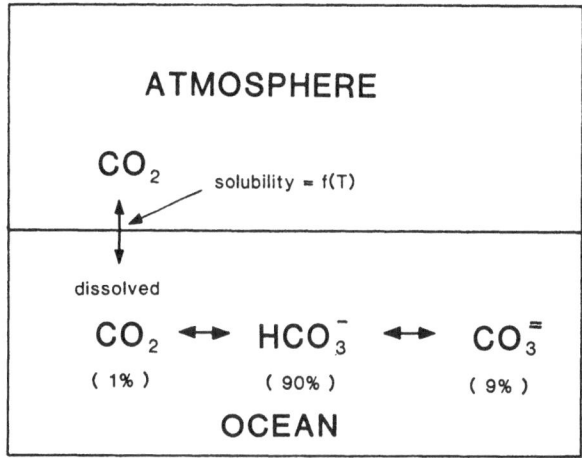

Figure 3: Schematic plot of the CO_2 exchange between air and sea. Total CO_2 (ΣCO_2) includes dissolved CO_2 gas, HCO_3^- and $CO_3^=$ ions.

Dissolved inorganic carbon or "total CO_2" (ΣCO_2) includes HCO_3^- ($\approx$ 90 %), $CO_3^=$ ($\approx$ 9 %) and dissolved CO_2 gas ($\approx$ 1 %; cf. Figure 3). At equilibrium between air and water, the concentrations of CO_2 in air, C_a, and in water, C_w, are proportional to each other:

$$C_w = s \ C_a \qquad (1)$$

s is the solubility. It decreases with increasing temperature, so lower temperatures during glacial times should have led, for constant C_w, to lower atmospheric CO_2 levels. During the glacial maximum 18,000

158

years ago, the average sea surface temperature was about 1.5 K lower
than now, which would have caused some CO_2 decrease. The combined effect of lower temperature and enhanced ocean salinity can be estimated
to have led to an atmospheric CO_2 reduction of only about 10 ppm, much
less than the observed 80 ppm change. Therefore, one has to look for
changes in C_w, i.e. in the oceanic carbonate chemistry, in order to
understand that change.

According to the relation (1), the atmospheric CO_2 level changes
proportionally to a change in the average concentration C_w of dissolved CO_2 gas in surface water. C_w mainly depends on ΣCO_2 in surface water; in addition, it depends on how the dissolved inorganic carbon is
partitioned between dissolved CO_2, HCO_3^- and $CO_3^=$. This partitioning
can be determined if another quantity, the alkalinity, is known and if
the equilibrium relations governing the aqueous carbonate chemistry
are considered. The alkalinity is approximately given by the sum of
the charges of bicarbonate and carbonate ions ($HCO_3^- + 2\ CO_3^=$); it is
affected by the formation and dissolution of carbonate shells. Here,
we will concentrate on the changes of ΣCO_2 that dominate other effects. A more detailed discussion of marine carbonate chemistry in
connection with atmospheric CO_2 can be found e.g. in Broecker and Peng
[1982] or in Siegenthaler [1986]. If ΣCO_2 increases by 1 % (with
constant alkalinity and temperature), C_w and consequently the atmospheric CO_2 concentration increases by roughly 10 %. Minor changes of
ΣCO_2 may therefore have considerable effect for the atmospheric composition.

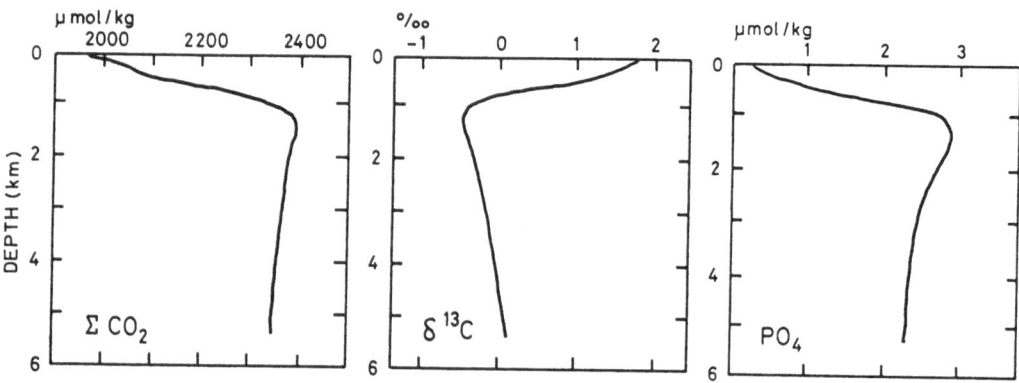

<u>Figure 4</u>: Depth distribution of ΣCO_2, $\delta^{13}C$ of ΣCO_2 and phosphate in
the mid-latitude North Pacific, after Broecker and Peng
(1982). The gradient between surface and deep water is caused by biological activity.

Now ΣCO_2 in surface water is smaller than for the average ocean (Figure 4), because dead organic particles and also carbonate shells formed by by small organisms sink from the surface to deeper layers, whre they are oxidized or dissolved. This mechanism is often termed the biological carbon pump. In average surface water, ΣCO_2 is 10-20 % lower than in deep sea water. The surface values, and therefore the atmospheric CO_2 concentration, are mainly regulated by the interplay of biological activity and transport by water circulation. In large oceanic areas, biological productivity is limited by the nutrients phosphate and nitrate, which are depleted there to nearly zero (cf. Figure 4). In these areas the biological pump strength is closely linked to water circulation: a change in the vertical circulation rate, and therefore in the supply of nutrients from deeper layers, involves a proportional change in biological activity, so that the chemical concentrations at the surface remain unaffected.

2.2   Possible causes of glacial-postglacial CO_2 changes

Broecker [1982] was the first who presented a coherent hypothesis for the CO_2 increase at the end of the last glaciation. He suggested that during glacial times, the average oceanic concentration of nutrients was considerably higher than during the Holocene, which would have led, for unchanged water circulation, to higher biological productivity and therefore to lower ΣCO_2 in surface water and a lower atmospheric CO_2 level. Broecker suggested that such a nutrient increase occurred when, at the beginning of glaciation, global sea level sank (because of formation of huge continental ice masses) and sediments rich in organic matter and nutrients from the freshly exposed continental shelves were transported to sea and redissolved. This hypothesis could explain the glacial-postglacial change in CO_2 and would also be compatible with changes of $\delta^{13}C$ observed in deep-sea sediments (see below). However, deposition and erosion of shelf sediments is a slow process and depends on major changes in sea level, while CO_2 changed within relatively short time. The shelf-sediment hypothesis is at contrast with recent evidence that the overall phosphate content of the ice-age ocean was presumably not much different than at present. This finding is based on the result of sediment studies that the average concentration of cadmium, which behaves very similarly as phosphorus, in the world ocean was about the same in glacial as in postglacial times [Boyle, 1987]. At the same time, however, these (and other) data clearly show that the ocean circulation pattern was different during glaciation than now, especially that formation of deep water in the North Atlantic Ocean, which is very important at present, was considerably reduced.

Another hypothesis is that atmospheric CO_2 reacted to changes in the high-latitude oceans. There the vertical water exchange is fast since it is not impeded by a thermal density stratification, and water rich in nutrients and CO_2 wells up from depth to the surface at a relatively high rate, so that the biological productivity is not high enough to consume all available nutrients (and the corresponding amounts of CO_2; cf. Figure 5). If the vertical water circulation would slown down then an unchanged biological productivity would lead to a stronger depletion of ΣCO_2 (and nutrients) in surface water, and consequently to an atmospheric CO_2 decrease. Model calculations show that a reduction of the vertical circulation in high latitudes by a factor two could have caused a CO_2 decrease by some 50 ppm [Knox and McElroy, 1984; Sarmiento and Toggweiler, 1984; Siegenthaler and Wenk, 1984].

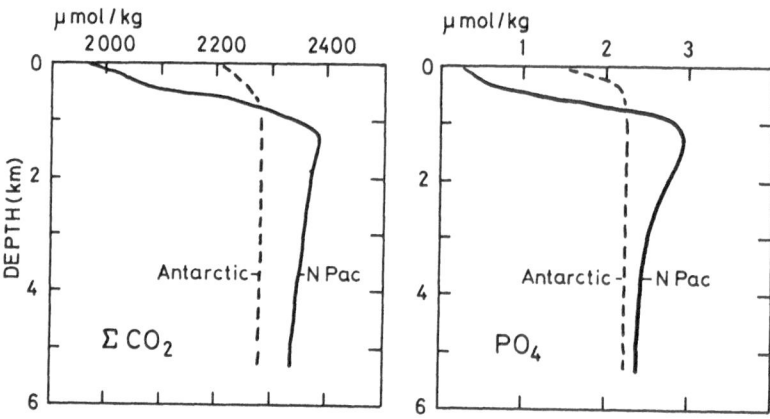

Figure 5: Typical vertical profiles of ΣCO_2 and phosphate in the mid-latitude North Pacific and in the Antarctic Ocean. In the Antarctic, phosphate concentrations and ΣCO_2 are relatively high at the surface.

As mentioned above, there is clear evidence from deep-sea sediment studies that the formation of North Atlantic Deep Water was indeed significantly reduced during the glacial time. This indeed could have affected atmospheric CO_2, according to the high-latitude - ocean hypothesis, in the direction observed; but the involved area in the North Atlantic Ocean seems to be too small to explain the whole effect. The Southern Ocean around Antarctica has the necessary large area and the required oceanographic properties, but there is no clear evidence for reduced vertical circulation in the Southern Ocean during glacial time.

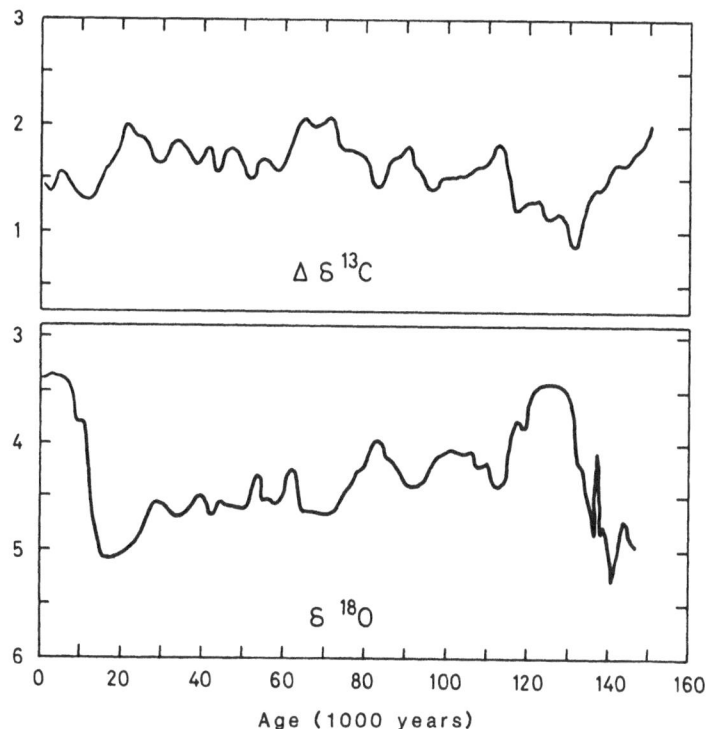

Figure 6: $\delta^{13}C$ difference between surface-dwelling and bottom-dwelling foraminifera, and $\delta^{18}O$ in bottom-dwelling foraminifera from a deep-sea sediment core from the eastern equatorial Pacific [Shackleton et al., 1983]. $\delta^{18}O$ essentially gives a record of global ice volume (high $\delta^{18}O$: large ice volume), because low-$\delta^{18}O$ water was transferred from ocean to land ice during glaciation.

The hypotheses about the cause of the CO_2 changes can in principle be checked by means of data on the isotope ratio $^{13}C/^{12}C$, since the influence of the biological pump is also reflected in the $^{13}C/^{12}C$ ratio of dissolved inorganic carbon (expressed by $\delta^{13}C$ = relative deviation of the $^{13}C/^{12}C$ ratio from that of a standard, in permil), as seen in Figure 4. Plants preferentially take up the lighter isotope ^{12}C during photosynthesis, and marine organic matter has $\delta^{13}C \approx -20$ %, compared to $\delta^{13}C \approx 0$ %o of total CO_2 in sea water. Consequently, the oxidation of organic particles leads to lower $\delta^{13}C$ at depth than in surface water. The difference in $\delta^{13}C$ between surface and deep water is roughly proportional to the ΣCO_2 difference. This fact may be used for estimating ΣCO_2 differences surface - depth in the past from $\delta^{13}C$ profiles measured in deep sea sediments. Shackleton et al. (1983) have found by $\delta^{13}C$ analyses on the carbonate shells of planktonic and ben-thonic foraminifera (tiny animals living in surface water and on the sea floor, respectively) of deep-sea sediments that the difference $\Delta\delta^{13}C$ between the two kinds of foraminifera was larger by about 0.5 ‰ during glacial time than during the Holocene and the last interglacial (Figure 6). This is exactly what is expected if indeed the biological pump mechanism is involved in the glacial-interglacial CO_2 variations. The $\Delta\delta^{13}C$ curve of Shackleton et al. varies nicely parallel to the CO_2 record from the ice core from Vostok, Antarctica, covering the past 160,000 years [Barnola et al., 1987], and the $\Delta\delta^{13}C$ variations agree in magnitude approximately with what the model calculations for the two mentioned hypotheses predict. According to the high-latitude - ocean hypothesis, $\delta^{13}C$ should have been higher in ice-age surface ocean water and in atmospheric CO_2. The available data on planktonic foraminifera seem not to confirm this expectation. $\delta^{13}C$ analyses on CO_2 trapped in ice-age ice should give the required information to check this reliably.

Thus, both existing model hypotheses about the cause of the gla-cial-interglacial CO_2 variations have their problems when compared with the existing data. While it seems rather likely that the CO_2 variations are connected to the biological carbon pump in the ocean, the precise mechanism is not yet fully understood.

3. Climatic effects of the CO_2 variations

It is now generally accepted that the cycle of ice ages and in-terglacials is connected to the quasi-periodic variations of the earth's orbital parameters (Milankovitch hypothesis). By spectral ana-lysis of climate records in deep-sea sediments and elsewhere, the same periodicities - ca. 21, 41 and 100 yr - are found as present in the orbital elements [Hays et al., 1976]. While the connection between as-tronomical variations and global climate appears well established, the mechanisms linking external cause and changes on earth are still far from well understood. The external forcing - changes in the seasonal insolation pattern - is, by itself, not strong enough to produce the ice ages; internal processes within the global climate must have strongly amplified them, and atmospheric carbon dioxide variations are such an internal process.

ENERGY BALANCE OF THE EARTH

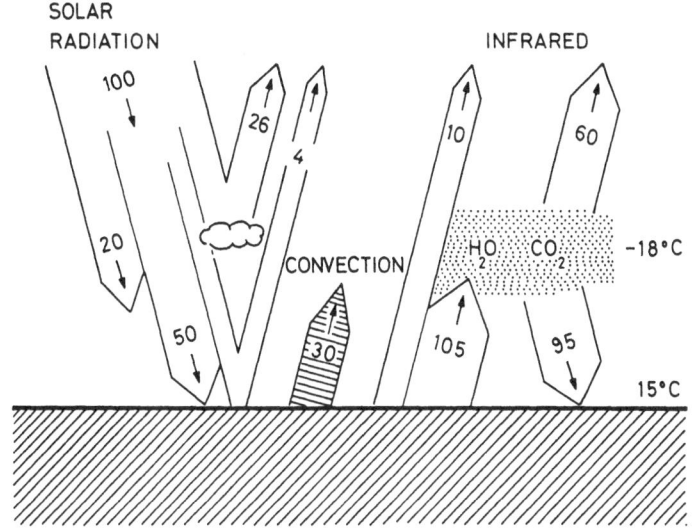

Figure 7: Energy balance of the earth. Incoming solar radiation = 100 units. Left side: short-wave length radiation; right side: infrared.

CO_2 influences the global climate via the greenhouse effect. Figure 7 shows the radiation balance of the earth. Of 100 units of incoming solar radiation, 30 are reflected back to space by clouds and the earth's surface. In a radiative equilibrium, a corresponding flux of energy is emitted from earth to space in the form of infrared radiation; the rate of infrared emission from the surface is proportional to the fourth power of its absolute temperature. The equilibrium temperature of the earth corresponding to the absorbed sunlight is 255 K or -18°C, i.e. considerably lower than the actual mean surface temperature of 15°C. The difference is due to energy transfer effects within the atmosphere, because water vapour, CO_2 and other gases are transparent for sunlight but not for the longer-wave length infrared radiation. Thus, most of the infrared emitted at the surface is absorbed within the atmosphere by these infrared-active gases, which re-emit upwards **and downwards.** In consequence, the infrared flux to space at the top of the atmosphere is equal to the net absorbed sunlight (70 units, cf. Figure 7), but the earth's surface receives additional 95 units of infrared back-radiation from the atmosphere. If the atmospheric concentration of CO_2 changes, then the greenhouse effect, and consequently the surface temperature, is modified.

Model calculations show that if the CO_2 level would double, then the mean surface temperature would rise by 1.2°C as a direct response. However, several feedback effects amplify the warming. A strong posi-tive feedback is due to the atmosphere's tendency to maintain a con-stant relative humidity, so that the absolute content in water vapour (which is also a greenhouse gas) increases as temperature goes up. Another feedback effect comes from the melting of snow and sea ice when the earth becomes warmer; the reduction of snow and ice cover leads to a decrease of the albedo (reflectivity for sunlight) of the surface. The overall mean temperature increase for a CO_2 doubling, as calculated by models of various complexity and for different model as-sumptions, is 2 to 4°C, with larger changes in high latitudes.

The question as to the contribution of the reduced CO_2 concentra-tion to the global cooling during the ice age, as simple as it sounds, cannot be answered in this form, because it is incomplete. It is ne-cessary to specify whether, and which, climatic feedback mechanisms are simultaneously changing, and additional information on ice-age conditions are necessary. Such information has become available from the CLIMAP (1976, 1981) project, which reconstructed, for the last glacial maximum, 18,000 years B.P. (before present), sea surface tem-peratures, extent of continental and sea ice, and surface albedo. The estimated mean sea surface temperature was 1.5°C colder than today; the global cooling, including continents, may have been about 4°C.

Two model studies of the climate 18,000 years B.P. are especially interesting and are now discussed [Hansen et al., 1984; Broccoli and Manabe, 1987]; both used general circulation models of the atmosphere, coupled to a mixed surface layer of the ocean (which is important for a realistic simulation of the seasonal temperature cycle). Hansen et al. run their model with the CLIMAP-reconstructed land and sea surface conditions as prescribed boundary conditions, while Broccoli and Mana-be only prescribed continental ice and land surface albedo, but not sea surface temperature which was thus simulated by the model and could be compared with the CLIMAP reconstruction. In both papers, the

sensitivity of the simulated climate to variations in individual factors, amongst other atmospheric CO_2, is studied.

Hansen et al. listed the contributions of individual changes
(without other feedbacks) to the overall difference in mean global
temperature between 18,000 years B.P. and today as follows:

Changed water vapour and clouds	1.4-2.2°C
Extent of continental ice	0.9°C
Extent of sea ice	0.7°C
CO_2 concentration (280 → 200 ppm)	0.5°C
Vegetation	0.3°C
Orbital variations	0±0.2°C

The total calculated cooling is 3.8 to 4.6°C. This value, as well
as the size of the individual contributions, clearly depends on the
specific model structure and assumptions.

The largest single contribution stems from changed water vapour
concentration and cloud cover. This is particularly interesting since
it is not a primary or external effect, but a purely internal feedback! The relatively large range has to do with uncertainty in the reaction of clouds, which is still an unsolved problem in climate modelling. The next important change is in the extent of continental ice,
which led to a considerably higher continental albedo during the ice
age. The change in type of vegetation affected the albedo of the land
areas not covered by ice, but with minor influence. The direct effect
of the CO_2 concentration change by a factor 1.4 amounts to only about
10 percent of the overall cooling; but a CO_2 decrease is automatically
amplified by changed water vapour and cloud cover which are fast feedbacks. Thus, a sudden change in CO_2 from 280 to 200 ppm, with present-
day continental and sea ice distribution, would induce a global mean
cooling by about 1°C. As a consequence, however, sea ice would
increase and therefore global albedo, again a positive feedback. On
the long run, also polar and continental ice might grow. In this way,
the different climate factors are interdependent and it is not always
possible to distinguish between cause and effect.

Broccoli and Manabe (1987) considered the influency of the following changes, separately and in combination: continental ice sheets,
CO_2, and albedo of non ice-covered land. Atmospheric water vapour and
sea ice were allowed to adjust themselves, i.e. acted as positive

feedbacks, but cloud cover was prescribed in a fixed way. The global mean cooling for the combined changes is 2.8°C, smaller than obtained by Hansen et al. (1984), perhaps because Broccoli and Manabe computed their mean only for regions free of continental ice. Broccoli and Manabe obtained the following temperature differences between ice-age and present conditions (influence of land albedo is omitted here, since rather small):

	Global	Northern Hemisphere	Southern Hemisphere
Extent of continental ice	1.3°	2.4°	0.3°
CO_2 concentration (300 → 200 ppm)	1.2°	1.1°	1.3°
Combined changes	2.8°	3.9°	1.9°

Thus, according to Broccoli and Manabe, continental ice sheets and changed CO_2 produce about the same global effect, while according to the results of Hansen et al., land ice seems somewhat more important; the results can, of course, not directly be compared, because Broccoli and Manabe include the effect of fast feedbacks. Most interesting is, however, the difference between the two hemispheres. The ice sheet effect is nearly fully restricted to the Northern Hemisphere, because there were no large ice sheets outside of Antarctica in the Southern Hemisphere; this result was also obtained by Hansen et al. (1984). On the other hand, the CO_2 decrease causes about the same cooling in both hemispheres. Broccoli and Manabe's results indicate that CO_2 was responsible for most of the cooling in the Southern Hemisphere.

This finding is most important in view of the fact that according to the available evidence, the glaciations seem to have been synchronous in the two hemispheres. This could not be expected a priori from the Milankovitch theory of the ice ages, since the seasonality of the insolation, which is supposed to be the ultimated driving mechanism, is not synchronous in the two hemispheres. A coupling mechanism between the two hemispheres is therefore needed, and according to the discussed climate model results, it might be found in the atmospheric CO_2, the concentration of which is homogeneous over both hemispheres.

4. A coupled global climate-CO_2 system

The climate model studies indicate that atmospheric CO_2 played a significant role for the glacial-interglacial cycle. On the other hand, as discussed in section 2, the current understanding is that varying conditions at the ocean surface, probably linked to climatic events, led to the observed variations in atmospheric CO_2 concentration. The question thus arises which was the hen and which the egg: did CO_2 change because of the modified climate and then acted as a feedback factor, or was the CO_2 change initiated directly by the Milankovitch-type insolation variations and then caused a shift in global climate?

At present, this question cannot yet be resolved with certainty. The ice core work, especially that by the French-Soviet collaboration on the Vostok core covering the past 160,000 years [Barnola et al., 1987], indicates that the CO_2 concentration varied remarkably parallel to climate. This demonstrates that both are intimately connected, but it does by itself not allow to decide about the cause-effect relationship.

One plausible hypothesis explaining the CO_2 changes by an external cause is that the seasonally varying insolation in the high latitudes might be involved. Knox and McElroy (1984) put forward the idea that e.g. during the last glacial maximum, high intensity of sunlight in early or late winter, i.e. a time when deep water is formed, might have resulted in higher biological activity and thus in lower CO_2 concentrations in surface water and in the atmosphere. It is difficult to test this hypothesis. While the idea is attractive that light-limitation may steer the biological productivity in high latitudes, one would not expect that the slow variation of the orbital parameters could be responsible for the CO_2 increase at the glacial-postglacial transition which fully took place within about 4000 years [Neftel et al., 1987]. If the rapid CO_2 variations observed during glacial time in Greenland ice [Stauffer et al., 1984] can be shown to be real, then the Milankovitch-type insolation cycles must clearly be ruled out as cause. In any case, it seems that oceanographic factors must have been involved, at least as strong feedback effects.

The current ideas are that insolation changes in high latitudes of the Northern Hemisphere were the main primary cause for the ice age cycles. There is clear evidence that formation of deep water in the

Northern Atlantic was reduced and different than today during glacial time (e.g. Boyle, 1987), which must have been connected to colder temperatures and considerably larger extent of sea ice. Conditions in the Southern Ocean around Antarctica, e.g. sea ice, certainly also changed, but there is no evidence for comparably drastic changes in deep water formation there. Thus it seems reasonable to assume that atmospheric CO_2 varied in response to the events in the North Atlantic Ocean. The problem here is that in the North Atlantic, the area of surface water in rapid contact with the deep sea seems to be too small to effect the atmospheric composition in the observed way, so that some indirect effects, e.g. a teleconnection to the Southern Ocean, must also be involved.

While it is not yet possible to give a consistent explanation of all events at the onset of an ice age, it is tempting to sketch a scenario. Figure 8 represents a personal view of how the combined climate and CO_2 changes might have arisen and be connected. According to this view, CO_2 changed in response to oceanographic events in the North Atlantic, induced by the Northern Hemisphere cooling and ice growth, and it then acted as a climatic feedback. The time delay between climate and CO_2 may have been rather small; as shown by Wenk and Siegenthaler (1984) the oceanic carbonate system adjusts itself within a few hundred years to changed circulation or productivity. The available CO_2 data from the glacial-postglacial transition do not allow to decide, within a few centuries, whether CO_2 preceded or followed climate.

Further modelling work is necessary to understand the CO_2 - climate system. Above all, however, more experimental data from ice cores and deep-sea sediments on the detailed history of CO_2 and its isotopes, nutrient concentrations in the oceans and of the ocean circulation are needed in order to gain more insight into these fascinating questions.

POSSIBLE CONNECTION CLIMATE − CO_2

TRANSITION   INTERGLACIAL → GLACIAL

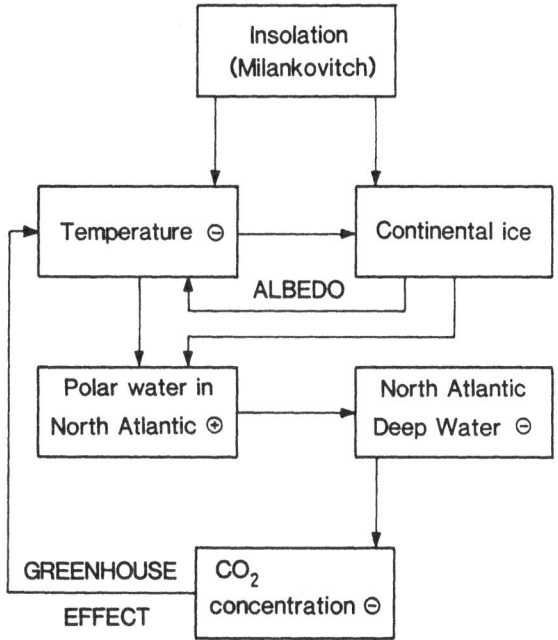

Figure 8: Scenario of the possible connection between processes in the combined climate-CO_2 system at the onset of an ice age. Insolation changes are the external driving force, all other compartments and processes belong to the interactive internal system.

REFERENCES

Barnola, J.M., Raynaud, D., Korotkevitch, Y.S., Lorius, C. 1987: Vostok ice core: a 160,000-year record of atmospheric CO_2. Nature, in press.

Boyle, E.A. 1987: Cadmium: chemical tracer of deep water paleoceanography. Paleoceanography, in press.

Broccoli, A.J., Manabe, S. 1987: The influence of continental ice, atmospheric CO_2, and land albedo on the climate of the last glacial maximum. Climate Dynamics 1, 87-99.

Broecker, W.S. 1982: Ocean chemistry during glacial time. Geochim. Cosmochim. Acta 46, 1689-1705.

Broecker, W.S., Peng, T.-H. 1982: Tracers in the Sea. Eldigio Press, Lamont-Doherty Geological Observatory, Palisades, NY 10964.

CLIMAP project members, 1976: The surface of the ice-age earth. Science 191, 1131-1137.

CLIMAP project members, 1981: Seasonal reconstruction of the earth's surface at the last glacial maximum. Geol. Soc. Am. Map Chart Ser. MC-36.

Delmas, R.J., Ascencio, J.M., Legrand, M. 1980: Polar ice evidence that atmospheric CO_2 20,000 y B.P. was 50 % of present. Nature 284, 155-157.

Hansen, J., Lacis, A., Rind, D., Russel, G., Stone, P., Fung, I., Ruedy, R., Lerner, J. 1984: Climate sensitivity: analysis of feedback effects. In: Climate Processes and Climate Sensitivity, Geophysical Monograph 29, 130-163. American Geophysical Union.

Hays, J.D., Imbrie, J., Shackleton, N.J. 1976: Variations in the earth's orbit: pacemaker of the ice age. Science 194, 1121-1132.

Knox, F., McElroy, M.B. 1984: Changes in atmospheric CO_2: Influences of the marine biota at high latitude. J. Geophys. Res. 89, 4629-4637.

Neftel, A., Oeschger, H., Schwander, J., Stauffer, B., Zumbrunn, R. 1982: Ice core sample measurements give atmospheric CO_2 content during the past 40,000 yr. Nature 295, 220-223.

Neftel, A., Oeschger, H., Staffelbach, T., Stauffer, B. 1987: CO_2 record in the Byrd ice core 50,000-5000 years B.P. Nature, in press.

Sarmiento, J.L., Toggweiler, J.R. 1984: A new model for the role of the oceans in determining atmospheric pCO_2. Nature 308, 621-624.

Shackleton, N.J., Hall, M.A., Line, J., Shuxi, C. 1983: Carbon isotope data in core V19-30 confirm reduced carbon dioxide concentrations in the ice age atmosphere. Nature 306, 319-322.

Siegenthaler, U. 1986: Carbon dioxide: its natural cycle and anthropogenic perturbation. In: Buat-Ménard, P. (ed.), The Role of Air-Sea Exchange in Geochemical Cycling, 209-247 (Reidel).

Siegenthaler, U., Wenk, T. 1984: Rapid atmospheric CO_2 variations and ocean circulation. Nature 308, 624-625.

Stauffer, B., Hofer, H., Oeschger, H., Schwander, J., Siegenthaler, U. 1984: Atmospheric CO_2 concentration during the last glaciation. Ann. Glaciol. 5, 160-164.

Stauffer, B., Neftel, A., Oeschger, H., Schwander, J. 1985: CO_2 concentration in air extracted from Greenland ice samples. In: Geophysical Monograph 33, 85-94, American Geophysical Union.

Stauffer, B., Oeschger, H. 1985: Gaseous components in the atmosphere and the historic record revealed by ice cores. Ann. Glaciol. 7, 54-59.

Wenk, T., Siegenthaler, U. 1984: The high-latitude ocean as a control of atmospheric CO_2. In: The Carbon Cycle and Atmospheric CO_2: Natural Variations Archean to Present, Geophysical Monograph 32, 185-194. American Geophysical Union.

Zumbrunn, R., Neftel, A., Oeschger, H. 1982: CO_2 measurements on 1-cm^3 ice samples with an IR laserspectrometer, Earth Planet. Sci. Let. 60, 318-324.

SUBJECT INDEX

ADDRESSES

A. Berger, Institut d'Astronomie et de Géophysique Georges Lemaitre, Université Catholique de Louvain, 2 Chemin du Cyclotron, B-1348 Louvain-la-Neuve, Belgium

P.L. Blanc, CEN/FAR, B.P. 6-92265 Fontenay aux Roses Cedex, France

J.C. Duplessey, Centre des Faibles Radioactivités, Laboratoire mixte CNRS-CEA, F-91190 Gif sur Yvette, France

C. Fröhlich, Physikalisch-Meteorologisches Observatorium, World Radiation Center, CH-7260 Davos Dorf, Switzerland

H. Grassl, Forschungszentrum Geesthacht, Max-Planck-Strasse 1, D -2054 Geesthacht, Federal Republic of Germany

P.D. Jones, Climatic Research Unit, School of Environmental Sciences, University of East Anglia, Norwich NR4 7TJ, United Kingdom

P.M. Kelly, Climatic Research Unit, School of Environmental Sciences, University of East Anglia, Norwich NR4 7TJ, United Kingdom

L. Labeyrie, Centre des Faibles Radioactivités, Laboratoire mixte CNRS-CEA, F-91190 Gif sur Yvette, France

C. Pfister, Historisches Institut, Universität Bern, Engehalden-strasse 4, CH-3012 Bern, Switzerland

F. Schweingruber, Swiss Federal Institute of Forestry Research, CH-8903 Birmensdorf, Switzerland

U. Siegenthaler, Physikalisches Institut, Universität Bern, Sidlerstrasse 5, CH- 3012 Bern, Switzerland

C. Tricot, Institut d'Astronomie et de Géophysique Georges Lemaitre, Université Catholique de Louvain, 2 Chemin du Cyclotron, B-1348 Louvain-la-Neuve, Belgium

H. Wanner, Geographisches Institut, Universität Bern, Hallerstrasse 12, CH-3012 Bern, Switzerland